The Lost Planets

The Lost Planets

PETER VAN DE KAMP AND THE VANISHING EXOPLANETS AROUND BARNARD'S STAR

John Wenz

Foreword by Corey S. Powell

THE MIT PRESS
CAMBRIDGE, MASSACHUSETTS
LONDON, ENGLAND

This book was set in Dante MT Pro and Heron Sans by Jen Jackowitz. Printed and bound in the United States of America.

Library of Congress Cataloging-in-Publication Data

Names: Wenz, John, author.
Title: The lost planets : Peter van de Kamp and the vanishing exoplanets around Barnard's Star / John Wenz ; foreword by Corey S. Powell.
Description: Cambridge, MA : The MIT Press, [2019] | Includes bibliographical references and index.
Identifiers: LCCN 2018058869 | ISBN 9780262042864 (hardcover : alk. paper)
Subjects: LCSH: Extrasolar planets. | Van de Kamp, Peter, 1901-1995. | Astronomy--History--20th century.
Classification: LCC QB820 .W46 2019 | DDC 520.92 [B] --dc23
LC record available at https://lccn.loc.gov/2018058869

10 9 8 7 6 5 4 3 2 1

Contents

Foreword

The Long Road to Many Worlds

Corey S. Powell

One of the greatest debates in the long history of astronomy has been that of exceptionalism versus mediocrity—and one of the great satisfactions of modern times has been watching the arguments for mediocrity emerge triumphant. Far more than just a high-minded clash of abstract ideas, this debate has shaped how we humans evaluate our place in the universe. It has defined, in important ways, how we measure the value of our existence.

In the scientific context, *exceptional* means something distinctly different than it does in the everyday language of, say, football commentary or restaurant reviews. To be exceptional is to be unique and solitary. To be mediocre is to be one of many, to be part of a community. If Earth is exceptional, then we might be profoundly alone. There might not be any other intelligent beings like ourselves in the universe. Perhaps no other habitable planets like ours. Perhaps no other planets at all, beyond the neighboring worlds of our own solar system.

If Earth is mediocre, the logic runs the other way. We might live in a galaxy teeming with planets, many of them potentially habitable, some of them actually harboring life. In the mediocre case, we little bipedal humans might not be the only sentient creatures peering out into the depths of space, wondering if anyone else is peering back.

Back in the 1940s, when Peter van de Kamp and Kaj Strand began searching in earnest for planets around other stars (what astronomers now call *exoplanets*), the issue of exceptionalism and mediocrity was still wide open. At the time, two ideas about the origin of our solar system were seriously considered, and they neatly landed on opposite sides of the divide.

According to the *nebular hypothesis*, proposed by French scholar Pierre-Simon Laplace in his 1796 treatise *Exposition du système du monde*, the Sun was born from a whirling cloud of gas. Earth and the rest of the solar system then formed from rings of gas thrown off by the Sun as it contracted under the pull of gravity. If other stars formed in the same way, then planetary systems like ours should be commonplace.

The *near-collision hypothesis*, advanced in 1917 by English physicist Sir James Jeans, pointed the opposite way. In this scenario, the Sun barely survived a close encounter with another star billions of years ago. The gravitational pull of the encounter drew a long filament of gas out of the Sun; that gas settled into orbit and eventually gave rise to the modern planets. (Jeans drew inspiration from Georges-Louis Leclerc, Comte de Buffon, who in 1749 suggested that the planets emerged from debris when a comet collided with the Sun. Buffon's model was the first physically motivated attempt to explain the origin of the solar system, and it nudged the early thinking toward the side of exceptionalism.)

Stars in the Milky Way are so far apart that close encounters should happen exceedingly rarely—about once every quadrillion years, based on modern calculations. If Jeans's predictions are correct, then our solar system might be unique, or at best there might be a single other set of planets circling the rogue star that was similarly tortured by the gravity of the Sun. In his 1922 Halley Lecture at Oxford University, Jeans proclaimed, "Astronomy does not know whether or not life is important in the scheme of things, but she begins to whisper that life must necessarily be somewhat rare."

Today, the broadest version of exceptionalism has been thoroughly disproved, largely by the painstaking work that John Wenz describes in this book. Equipped with vastly superior instruments, van de Kamp's successors have discovered 3,944 confirmed exoplanets by the most recent count, and the tally increases almost daily. The roster of alien worlds includes a remarkable variety of forms, many of which have no equivalent in our solar system.

Astronomers do not yet have the technology needed to find a close analog of Earth orbiting a close analog of the Sun, so we still know little about how common or rare such worlds may be. The question of alien life is still wide open. What we do know is that the Milky Way is home to a tremendous number of other planets, probably trillions of them. In that sense, at least, we are certainly not exceptional, and Earth is certainly not alone.

The notion of cosmic mediocrity that inspired van de Kamp and the other early planet hunters is so old that it predates modern observatories. It predates the seventeenth-century invention of the telescope. It predates even what could recognizably be called "science" in the modern sense, tracing its origins at least back to the Greek philosopher Anaxagoras of Clazomenae, who wrote and taught in Athens in the fifth century BCE.

Anaxagoras proposed that the cosmos is ruled by an all-pervasive intellect, which he called *nous*, and that this intellect functions as a set of universal laws—a philosophical ancestor of Isaac Newton's theory of universal gravitation. Under the action of *nous*, the elements of nature were set into circular motion, separating into different components. The Sun, a ball of incendiary metal, was cast off into the sky by this process. So, too, were the stars and planets. Although what survives of Anaxagoras's writing is fragmentary and mostly secondhand, he seems to have imagined the stars to be fiery lumps much like the Sun, just drastically more distant. In one especially intriguing passage, he further hints at the existence of other lands similar to Earth and expansively argues "that there are a sun and a moon and other heavenly bodies for them, just as with us."

Many of these ideas reappeared in a philosophy even more ahead of its time, that of Aristarchus of Samos. During the third century BCE, Aristarchus advanced the first known heliocentric model of the solar system, evicting the Earth from its long-assumed central position and completely reworking the order of the cosmos. There is no surviving description of this iconoclastic model in Aristarchus's own words. Fortunately, his contemporary Archimedes provided a succinct summary:

> His hypotheses are that the fixed stars and the Sun remain unmoved, that the Earth revolves about the Sun in the circumference of a circle, the Sun lying in the middle of the orbit, and that the sphere of the fixed stars, situated about the same center as the Sun, is so great that the circle in which he supposes the Earth to revolve bears such a proportion to the distance of the fixed stars as the center of the sphere bears to its surface.

That final idea, though somewhat obscure in its phrasing, is pregnant with significance. Aristarchus is saying that the stars are so far away that we cannot see their *parallax*: They appear stationary even as the Earth moves in a great circle around the Sun. The implications are twofold. First, he imagined a cosmos vastly larger than the one implied by the geocentric system. Second, he reiterated and expanded on Anaxagoras's deduction that the stars might be other suns, this time explicitly spelling out the kinds of grand distances necessary for the stars to nevertheless appear as fixed cold dots in our sky.

The budding possibility of a multitude of worlds fully blossomed in the philosophy of the Greek atomists, most notably Epicurus. They envisioned not just other stars but other entire *kosmoi* (cosmic systems) beyond the one we know, each following the inexorable rules of the atoms it contains. Writing at about the same time as Aristarchus, Epicurus declared, "There is an infinite number of worlds, some like this world, others unlike it. For the atoms being infinite in number . . . are borne ever farther in their course." His atoms were mathematical and ethical constructs, quite unlike the physically described quantum units of today's physics, and yet, in the way Epicurus reached toward a boundless universe, he sounds shockingly prescient.

That pinnacle of glorious Epicurean mediocrity, alas, was followed by a lengthy retreat into a constricted, Earth-centered cosmology. Aristotle retorted, "There cannot be more worlds than one," and his great authority carried the day. Around 150 CE, Claudius Ptolemy shrank even further from the *kosmoi* when he merged Aristotelian physics with state-of-the-art observations of stars and planets into a unified Earth-centered model. The Ptolemaic system consisted of a set of nested celestial spheres, dispensing with exotic speculations about infinite space and other suns. By Ptolemy's reckoning, the outermost crystalline sphere, containing the fixed stars, was about 20,000 times the radius of the Earth, making his entire cosmos just 160,000 miles wide in modern terms.

What the Ptolemaic system lacked in grandeur, it made up in practicality. It predicted the motions of the planets and stars with admirable precision using a combination of mathematically appealing circular motions. Ptolemy's astronomical writings, later translated by medieval Islamic scholars as the *Almagest* (literally "the greatest" in Arabic), reigned supreme for more than a millennium. His authority was cemented when prominent theologians like Thomas Aquinas merged the Ptolemaic system with the Roman Catholic worldview during the Middle Ages. The outermost sphere of the cosmos equated with heaven, and the Aristotelian "prime mover," who set the spheres in motion, became one and the same as the Christian God.

The attributes that make exceptionalism appear impoverished from a scientific perspective make it precious from a theological point of view: only one Earth, one heaven, one God. But the fire of human imagination is not so easily snuffed. Some medieval Islamic astronomers continued to speculate about the existence of other worlds. Catholic scholars, too, pushed against the boundaries. Around 1450—a full century before the mystical speculations of Giordano Bruno—the German philosopher and astronomer Nicholas of Cusa wrote about the notion of infinite space, in contradiction to Ptolemaic concepts. Nicholas structured his ideas within a Catholic framework, exploring infinity as a natural corollary to the limitless glory of God, but his philosophy kept alive the possibility of a physically unbounded universe as well.

Then along came Nicolaus Copernicus, and mediocrity began a full-on comeback.

From outward appearances, Copernicus was an unlikely figure to knock the solar system askew and to set astronomy on its modern path toward a multitude of planets. He worked as a canon in Warmia, a small, semiautonomous Catholic state in what is now Poland, tending to various local political and economic disputes. He was a modest, well-liked figure, not particularly known for his controversial opinions. Professionally, his most notable achievements were probably in economics and monetary theory. There was a spark within that set him apart, however: the bold, revisionist astronomical ideas brewing in his head.

Sometime before 1514, while he was still in his thirties, Copernicus wrote a summary of his new model of the solar system. Influenced by the arguments of Aristarchus, as well as by his own strong sense of the Ptolemaic system's mathematical ugliness, Copernicus returned the Sun to the center and set the Earth in motion around it. He circulated his short document, called the Commentariolus, among his friends, with the intention of expanding its arguments into a fully developed work of heliocentric cosmology. That magnum opus, *De revolutionibus orbium coelestium* (*On the Revolutions of the Heavenly Spheres*), was famously not published until 1543, when he was on his deathbed. Copernicus was unconscious when a finished copy was thrust into his limp hands, and he died that same day.

The publication delay was not, as popular accounts often claim, a simple matter of Copernicus's fear of the Catholic Church. He was more afraid of the Church's intellectual partners, the Aristotelian philosophers, who he worried (not unreasonably) might be brutal to this upstart living far from the intellectual heart of Europe. He also needed to perform detailed mathematical analysis and to collect astronomical observations in support of a theory that he was developing in his spare time. Only in retrospect do those fears look absurd. It turns out that the time was ripe for a critical reexamination of entrenched classical Greek thinking. In the decades after its publication, *De revolutionibus* was extensively read and discussed across Europe. The influential Danish astronomer Tycho Brahe even praised Copernicus as "a second Ptolemy."

Two disciples of Copernicus were especially pivotal in establishing Copernican mediocrity—the notion that Earth does not sit in a privileged position but is representative of the richness of the universe as a whole. In 1575, Thomas Digges, a leading astronomer in sixteenth-century England, published the first English translation of *De revolutionibus*. He added commentary to clarify that the Copernican system was a physically realistic model of the solar system (not just a computational trick), and he overtly broached the idea that a sun-centered universe could be infinite in extent. To drive home this last point, Digges created a drawing showing, for the first time in history, how the stars might be scattered through endless space outside our solar system.

A few years later, the German astronomer Michael Maestlin adopted Copernicus's system as superior to Ptolemy's and spread heliocentric thinking broadly from his prominent position as a teacher at the University of Tübingen. Most notable among his students was a clever young fellow named Johannes Kepler, who, starting in 1609, figured out that planets go around the Sun in elliptical paths. This discovery thoroughly and finally smashed Ptolemy's claustrophobic crystalline spheres. The universe was now wide open to all possibilities, and to endless worlds.

I won't belabor here the too-familiar stories about Galileo's tussles with the Church, or Isaac Newton's development of the laws of universal gravitation, except to note how rapidly (historically speaking) the concepts of Copernican mediocrity spread and triumphed. By the middle of the seventeenth century, heliocentrism was widely accepted across the Western world. By the eighteenth century, many leading intellectuals embraced not only the idea of other worlds, but even other *inhabited* worlds. Cyrano de Bergerac's *Comical History of the States and Empires of the Moon*, published in 1657, introduced the reader to imaginary inhabitants of the Moon. Jonathan Swift's *Gulliver's Travels* (1726) and Voltaire's *Micromégas* (1752), whose central character is a being from a planet orbiting the star Sirius, casually assume a multiplicity of inhabited worlds as a backdrop to their social satire. Mediocrity was in vogue.

On the scientific side, William Herschel (perhaps the most famous astronomer of the late eighteenth and early nineteenth century) was a firm proponent of the idea that life is common on other planets. Right around the time that he discovered the planet Uranus, Herschel shared what he believed to be telescopic evidence of intelligent life on the Moon. He later argued that all worlds might be inhabited; improbable as it sounds, he even suggested that life exists on the Sun, huddled beneath the luminous clouds covering its surface. Although many other researchers were not so enthusiastic, each generation found its champion of life beyond Earth. American astronomer Percival Lowell was especially effective at promoting such ideas well into the twentieth century with his popular (if increasingly eccentric) writings about an imperiled advanced civilization on Mars.

Science fiction writers like Ray Bradbury, Arthur C. Clarke, Isaac Asimov, and Robert A. Heinlein further popularized many-worlds mediocrity with their compelling visions of alien inhabitants on far-off worlds. So did the pulp science fiction and fantasy comic books of the post–World War II era. Although reputable scientists looked askance at most of this wishful thinking, mainstream astronomers were still routinely speculating about jungles on Venus and creeping lichens on Mars at the time workers at Sproul Observatory were setting their sights on planets around other stars. It was an optimism that would soon collapse.

Just as astronomers were developing the tools needed to convincingly detect planets around other stars, aerospace engineers were refining the rockets and spacecraft that would soon shred any confidence that life exists closer to home, elsewhere in the solar system. Although the emerging fields of exoplanet research and astrobiology were not directly related, they were inextricably linked. Dashed expectations for finding life nearby inevitably dashed hopes for finding life around other stars, stirring skepticism about the reality of those planets. The disheartening controversies over claimed planets around 61 Cygni, Barnard's Star, and other nearby stars further soured the opinion of the scientific community.

Early findings from US and Soviet space probes almost uniformly made the solar system seem shockingly hostile to life. NASA's Mariner

2 flew past Venus in 1962 and found that the planet is not a steamy jungle at all; rather, it is an Earth-size sterilizing oven, with a crushing atmosphere and surface temperatures hovering around 800 degrees Fahrenheit. Two years later, Mariner 4 visited Mars and beamed back images of a barren cratered landscape to crestfallen planetary scientists. In 1976, the year that van de Kamp retired from Sproul Observatory, NASA sent the twin *Viking* landers to Mars to do a Hail Mary search for life right there on the surface. The $1 billion effort, equivalent to $5 billion today, yielded no conclusive signs of anything alive.

Pioneering astrobiologists like Carl Sagan realized that life on nearby worlds, if it exists, would be much more subtle and challenging to detect than astronomers had originally hoped. Likewise, as John Wenz details in the pages ahead, the many ambiguous detections and retracted claims from astrometric searches for exoplanets cast doubt on the feasibility of the whole endeavor.

By this point, the few stalwarts in the field understood that finding planets would require new techniques—but who would devote the necessary time and energy to advancing a field that had become so disreputable? In 2012, Gordon Walker at the University of British Columbia, one of the founders of the radial velocity search technique, published a scientific memoir, "The First High-Precision Radial Velocity Search for Extra Solar Planets," in *New Astronomy Reviews*, looking back to those early days. "It is quite hard nowadays to realize the atmosphere of skepticism and indifference in the 1980s to proposed searches for extra-solar planets. Some people felt that such an undertaking was not even a legitimate part of astronomy," he wrote.

Conducting research under those conditions was an exercise in masochism. "We applied twice each year for telescope time and were generally assigned four pairs of nights per year—although one year we received none in the first six months," Walker continued. "It really was tedious because there could be no obvious results from any one observing run and, perhaps more seriously, no publications to nourish research funds. Nonetheless, we did persist for 12 years and, it seems, made some exoplanetary discoveries in the process—but it took some time for these to dawn, because none was a simple case."

Walker at least managed to persist in his academic astronomy career. His graduate student Bruce Campbell—who did more than anyone else to make radial velocity a viable way to find exoplanets—did not fare as well. John Wenz's description of this episode stands as one of the book's more tragic passages. Campbell's bitter departure from the world of research has many echoes in history. From our time-compressed modern perspective, we look back at Copernicus as a triumphant figure, but he spent more than three decades of his life working on astronomy as an unpaid sideline. He didn't live to see his summary work published. Even after that, his ideas were not widely accepted for another century.

Perhaps someday future historians of science will similarly reflect on Walker and Campbell as unsung heroes of our current era. Then again, we don't have to wait for some imagined far-off historical verdict. We know right now that Walker and Campbell were on the cusp of momentous discoveries. Their exact technique, deployed in a slightly different way, led to the first clear-cut detections of planets around Sunlike stars beginning in 1995. Those detections, in turn, provided the scientific confidence needed to gain approval for NASA's big-budget Kepler space telescope and its successors, including the brand-new TESS (Transiting Exoplanet Survey Satellite). Walker and Campbell, like van de Kamp before them, were pioneers of mediocrity.

Bolstered by the bounty of discoveries from current exoplanet surveys, mediocrity is ascendant on the astrobiology side as well. Fears that Earth might be the single instance of a living planet seem much less plausible—if not downright unlikely—when researchers start to consider not just the other worlds of our solar system, but the trillions of other planets scattered across our galaxy. Sara Seager of MIT, the deputy director of the TESS mission, has set a lifetime goal of finding 500 planets similar to Earth. "If we're lucky, maybe 100 of them will show biosignatures," she says, referring to the data readings that would indicate the presence of life. Thirty years ago, such a statement would have provoked derisive laughter, or worse.

Finding 100 living planets would be amazing. With a sample that size, scientists could compare the different types of environments that support life, different styles of metabolism, and different stages of evolution. They could navigate to whole new levels of mediocrity, exploring Earth's place within an entire pantheon of inhabited worlds. Even a *single* living world would be a breakthrough, an unprecedented connection between humanity and the rest of the universe. There is no way to know when such a discovery will happen. There's no way to be certain it will happen at all. But thanks to the hunters of worlds lost and found—the imperturbable obsessives like van de Kamp, Strand, Walker, and Campbell—the possibility is wide open before us.

Acknowledgments

I am grateful to my editor Jermey Matthews, who guided me through the process of completing my first book. I would additionally like to thank Corey S. Powell, whose curiosity and body of work I hold in high esteem, for agreeing to write the foreword for the book. I'd also like to thank Virginia Crossman and Melanie Mallon for their editing guidance and patience. My acknowledgments would not be complete without mentioning the work of Christine Bridget Savage and Gabriela Bueno Gibbs, both an integral part of my MIT Press experience.

I would also like to thank Lisa Ruth Rand for her love and support throughout the process of completing the book, a path that wasn't always easy, personally or professionally. She also guided me through the academic processes of putting together my research, an intellectual asset that must be acknowledged as well. I would also like to thank my family—my parents, siblings, nieces and nephews, aunts, uncles, cousins, and more for their support. There are too many of them to contain in a small list, but each has been a meaningful part of making this book come together. In addition to my blood family, there's of course a long list of friends to acknowledge, several of whom were helpful during my book research phase. Specifically, I'd have been nowhere without the hospitality of my dear friend Laura Chance, who housed me while I researched my book at Swarthmore, as did Joseph Malcomson and Marissa Nicosia. Jen Barnason, Chris Vogt, Karl Spurzem, and so many others were there to guide me through the process as well.

The staff of the Friends Historical Library of Swarthmore College, especially Celia Caust-Ellenbogen, were incredibly helpful and patient as I researched my book at their institution and they guided me through the archives of the Sproul Observatory. I would also like to thank Eric Jensen, Swarthmore professor of astronomy, for showing me the modern-day facilities and speaking with me about the book.

I'd like to thank Ignasi Ribas at Institut de Ciències de l'Espai, Guillem Anglada-Escude at Instituto de Astrofísica de Andalucía, and Abel Mendez at the Planetary Habitability Laboratory at the University of Puerto Rico at Arecibo for enlightening me on their research into Barnard's Star and allowing me access to their research before its official publication.

The maddening process of putting together a book would have been more difficult without the support of several colleagues whom I would like to thank. I'm especially grateful to Shannon Stirone, Adam Becker, Sarah Scoles, Joshua Sokol, and Mika McKinnon, who all provided invaluable guidance.

The staffs of *Popular Mechanics*, *Astronomy* magazine, and *Discover* magazine have been spectacular colleagues whom I consider friends. I would especially like to thank Andrew Moseman and Eric Limer at *Popular Mechanics*; Kathi Kube, Rich Talcott, Michael Bakich, David Eicher, and Alison Klesman at *Astronomy* magazine; and Eric Betz, Nathaniel Scharpling, and Bill Andrews at *Discover* magazine. I have also had the fortune of working with other tremendous editors, including Jay Bennett at *Smithsonian* magazine.

Introduction

In 1600, Giordano Bruno was burned at the stake for his radical views—that not only was the Sun just one of many stars, but those stars likely had planets around them as well.[1]

Today, we seem to be approaching *innumerable*. Whereas a single planet discovery used to make headlines, now the planet has to have something special about it to really draw anyone's attention: the closest, the weirdest, the most habitable, the . . . somethingest. And even then, there's fatigue over the latest habitable exoplanet (meaning a planet outside our solar system). In some ways, the Kepler spacecraft's massive data dump on exoplanets answered the question "how many stars have planets" with "seemingly all—or most—of them," while dropping so many discoveries that it became impossible to keep up. The Kepler findings work better as a census than insight into individual worlds.

Bruno's proclamation has ended up seeming to be the case. We're just now catching our breath from that realization, ready to settle down and confront what it actually means.

We've only ever spotted a few planets directly—always young, always hot from formation, always large like Jupiter, and always just a small cluster of pixels. Planets are drowned out by the light of their stars, so we typically use indirect clues to find them, as the Kepler craft did, waiting for planets to pass in front of their stars and cause nearly imperceptible drops in the flux of light coming from the star. Or even more indirect methods, like watching how a planet tugs on its star, moving it ever so slightly, or waiting for a star to act like a giant magnifying glass on an object behind it, making evidence for a planet easier to find, at least temporarily.

The 2020s will give us some of our first true glimpses of planets, not as specks of light or as shadowy moving ghosts but as living, breathing worlds as beautiful and varied as those in our own solar system. The biggest telescopes in history will be open for business by 2030, like the 129-foot European Extremely Large Telescope in 2024, and the 80-foot Giant Magellan Telescope in 2021. There's also the Thirty Meter Telescope, which will have a 98-foot mirror should it ever break ground. (The project was delayed by protests from Native Hawaiians over the use of sacred land.) That's not to mention NASA's long-delayed Hubble successor, the James Webb Space Telescope. The massive 69.5-by-46.5-foot telescope will move to an orbit beyond the moon in 2021, open up its fairings, and begin to look at nearby planets with gusto.

These telescopes have plenty of tantalizing targets to pursue: There's Proxima Centauri b, the closest planet to our solar system for the next 36,000 years. There's the TRAPPIST-1 system, a planetary system encompassing seven Earth- and Mars-sized planets orbiting a star barely bigger than Jupiter, and at least three of those planets are potentially habitable, if not all seven. There's Luyten's Star, where Search for Extraterrestrial Intelligence (SETI) researchers recently beamed a message from Earth that should reach that system in 2029. The Tau Ceti system will likely be considered. Or a new planet may come shining out of the data from the Transiting Exoplanet Survey Satellite (TESS), an MIT and NASA collaboration to look for transiting planets around stars within 300 light years of Earth. The mission was designed to find the most ideal planets for the Webb telescope to examine.

In other words, a new era of planetary discovery just over the horizon is going to clarify what other planets look like—and whether life is a fluke of Earth or a widespread phenomenon.

The knowledge we seek is the kind of thing that, once upon a time, would have gotten you burned at the stake. But in between Bruno and Webb, we accumulated a serious bit of knowledge for understanding our universe. Not only did we come to realize that the Earth revolves around the Sun, but we realized that those other points of light out there were the same things as the Sun—burning cauldrons known as stars. And our own Sun drags its planets, comets, asteroids, and whatever we're calling Pluto now around the center of our galaxy. Even understanding that there is such a thing as a galaxy was a major coup.

This trajectory is rough—and obviously heavily condensed. But even though plenty of planets exist outside our solar system in science fiction, the number in science *fact* in the early twentieth century was 0. But astronomers were beginning to discover a greater number of small red stars closer to us than they'd ever anticipated, and a few of these are orbiting other stars. Indeed, to this day, we're *still* discovering faint neighboring stars and almost-stars.

The nineteenth century brought a new fold into astronomy, first in 1862, and then in 1895—that there are stars we can't see but can infer from the way they move the stars they orbit off a common center point. The first such discovery was the star Sirius, the brightest in the night sky. It had concealed a hidden companion, called Sirius B, a *degenerate* star. The second was the companion to Procyon, a bright star in the constellation Canis Minor. Look to the shoulder of Orion—one of the most easily recognizable constellations in the night sky—and move to your "left," toward a bright star. That star is Procyon. Trace down at roughly a 45-degree angle and you'll find Sirius, which is hard to miss because of its brightness. The star in the shoulder of Orion is known as Betelgeuse, and with Procyon and Sirius, it forms the *winter triangle*. Betelgeuse, unlike its triangular pals, seemingly has no "invisible" companion.

These companions also aren't exactly invisible. They are simply much dimmer than their parent star. But the method of discovering them was giving astronomers food for thought. Binary stars were nothing new. The star Algol was known to vary in brightness as a result of a companion star continually eclipsing it in the night sky, giving rise to the astronomical term *Algol variable*. William Herschel—the man who presented Uranus to the world in 1781—also systemically hunted down and cataloged binary stars. But the idea that one of these stars could be "invisible" yet still produce a pronounced effect on its parent star meant that plenty of invisible binary objects could be out there.

It also meant that planets would have this same effect, just . . . smaller. If we could find very small stars, it was reasoned, then we could find something even smaller—planets, whole worlds—despite the vast distances. Some might be as inhospitable as Neptune or Jupiter, but others could be a lot more like Earth.

Planets don't orbit stars in a neat and orderly fashion. Nor do moons orbit planets in a circle. Instead, the tug-of-war between the two gives both objects a common center of mass, called the *barycenter*. While the Moon orbits Earth, it exerts enough of a gravitational tug that the common center of mass is far past the Earth's core and well into the mantle. The common point of gravity between Jupiter and the Sun is actually outside the Sun, because both bodies have enough gravitational pull to hold their own against another object.

If an alien civilization looked in on the Sun with the tools we had at our disposal when planet hunting began in earnest, they would likely detect only Jupiter and maybe Saturn. They wouldn't be able to "see" Jupiter either. They'd instead focus an instrument on the Sun that would notice how much Jupiter pulled on it from their common point of gravity.

Night after night, astronomers were taking pictures of the stars in the night sky. Perhaps, they reasoned, we would see a star deviate ever so slightly from its path, as we saw with Sirius and Procyon, but on an even smaller scale.

Plenty of observatories were on the hunt for invisible stars. But when Dutch astronomer Peter van de Kamp came to Swarthmore in 1937, he was on the hunt for something else: invisible planets.

It would both make and break his career in the coming decades.

1
2
3
4
5
6
7
8
9

In the midst of war tearing one world apart, another was seemingly discovered.

In 1942, US Air Force private Kaj Strand was training for war at Eglin Air Field in Florida. He undertook tests on the B-29 Superfortress, one of the most fearsome weapons of the war that the United States had been thrust into. As soon as he gained citizenship, Strand intended to do his part for the war effort. He and fellow astrophysicist Martin Schwarzschild had "both wanted to fight Hitler." Strand was 35 at the time but ready to serve his adopted country. In correspondence back home, however, he had another matter to attend to.

After five years of work, Strand's close observations of the star 61 Cygni were ready to be presented to the world, and he had something audacious to announce: the first ever planet discovered outside our solar system.

Strand hadn't seen it, at least not any more directly than we have seen planets since. Instead, like the thousands of exoplanets discovered in recent decades, Strand's discovery relied on indirect methods. One such method, which Strand used in monitoring 61 Cygni, is known as *astrometry*. Astrometry, at its most fundamental level, is the study of the position of stars in relation to one another. Distance in space is measured in light years, which is the amount of time it takes light emitted by one object to reach another. One light year is 5.88 trillion miles. The star Sirius A is 8.6 light years away. When we look at it in the night sky, we're seeing the star as it was nearly nine years ago, all thanks to the funny workings of physics. Sirius is the brightest star in the night sky.

Altair is another nearby bright star, 16 light years away. Yet its distance from Sirius is 25 light years because Altair is on the other side of Earth, and stars sit in every direction from us—up, down, sideways, and at all angles. The way we can gauge these distances, and map the positions of the stars, is by closely monitoring how they move across the night sky, in a dance called the *proper motion* of the star. Some call this triangulation of positions *parallax*. We see how one star moves in comparison to stars that are more distant. Changes in the motion of the stars indicate that something is tugging on that star.

Sproul Observatory at Swarthmore College, in a sleepy suburb of Philadelphia, was gaining a reputation for its study of astrometry in the years just before the war. Observatory director Peter van de Kamp (Strand's advisor) and his associates, adjunct researchers, and underlings were focusing their time with the 24-inch telescope on studying the proper motion of most nearby stars. Sproul was already discovering invisible stars, binary ones that were detectable only by the influence they exerted on their primary star, rather than by their own light. A smaller star might reveal itself by causing its primary to deviate off an otherwise smooth pass across the sky. These motions were nearly undetectable except by the keenest observer. Some of these deviations required intense scrutiny and hundreds of observations.

Careful, repeated observation has always been a core part of astronomy. In 190 BC, Hipparchus drew up the positions of 850 stars, long before astronomers had telescopes at their disposal. He also discovered the position of Earth in relation to the Sun, all by watching how stars moved in their seasonal paths. The first star discovered by astrometry was a tiny companion to the bright Dog Star, Sirius. It was predicted to exist in 1830 based on Sirius's odd undulating movements, and it was visually confirmed in 1862. Astrometry is still used today, most specifically on the European Space Agency's Gaia mission, which is mapping the precise positions of stars—and waiting to see if they move at all from a center line, indicating invisible companions.

Usually, two stars in a gravitational embrace have a profound effect on the other's orbit. Although "easy" would be a misnomer, a star tugging on another star is one of the easiest invisible companions to detect. But planets also influence the motions of their star, albeit in smaller ways, because of the common point of gravity, or barycenter, they share. The Earth and the Moon are locked in such an embrace. Because the Moon has a substantial amount of mass, the barycenter is in the Earth about 2,900 miles (4,670 km) away from the planet's core—firmly within the Earth's mantle, but nearer to the surface than the core.

The barycenter of the Earth-Moon system in relation to the Sun is outside the Sun's core, which is impressive considering that the Sun has the mass of 333,060 Earths. But more impressive is that Jupiter, despite having maybe 1 percent the mass of the Sun, has a barycenter in space outside the star. This means that Jupiter causes the Sun to move in such a way that alien observers, with the right equipment, could watch the invisible tug of the Sun by Jupiter. This tug may not be as pronounced as that of a star-sized binary would be, but it's there.

Strand, with van de Kamp's guidance, had uncovered one of the smallest changes in the motion of a star. At the time, astronomers already knew that the 61 Cygni system had two stars, but Strand inferred that one of those stars had a fairly small object tugging on it as well, something with 16 times the mass of Jupiter. In short, Strand thought he was seeing an enormous gas giant, all just 11 light years away from us, a hop, skip, and a jump in our tens-of-billions-of-light-year spanning universe.

In February 1943, Strand's paper was officially published, announcing to the world the existence of the small companion.[1] Just a month before, van de Kamp had been drumming up publicity for the planet in the pages of the *New York Times*. "It is likely that an object of such a small mass has no luminosity of its own, and may therefore be tentatively classified as a planet, rather than a star," van de Kamp wrote to William Leonard Laurence, the then-leading science reporter for the *New York Times*.[2] More details emerged in the January 1943 issue of *Sky and Telescope*, which proclaimed in its news section that the planet had a five-year orbit and that it didn't so much circle around the star as loop around it in an oval, which brought it to within 65 million miles of its star before swooping back out.[3]

Today, finding a planet is commonplace. There's a new announcement seemingly every week, and that's only for the worlds that stand out. With around 4,000 known exoplanets out there, some humdrum Jupiters slip past our notice in favor of gleaming, maybe-habitable Earth-size worlds or star-scorched hell worlds that rain diamonds and produce the same chemicals as sunscreen.

Soon after Strand published, somebody impersonating a relative of his told the *New York Times* that the planet had a name. Had the person behind the alias Graham Strand gotten his way, the planet would have gone down in the history books as Osiris.

The faux Strand had told the *New York Sun* a similar tale, with that paper launching an investigation into the matter. Kaj Strand—the real one—told the *Sun*, "I have not thought seriously of naming that planet," with further attempts to find the Grand Graham Fraud hitting a dead end.[4] Author James Hickey concluded, "The mystery of who Graham R. Strand is and what he was up to in sending out the false publicity remains as far from a solution as ever. Not even the stars can tell you what he had in mind." In a letter to van de Kamp, he speculated that "he may be an educated person with a screw loose."[5]

It was a bizarre start to the exoplanet bonanza that would ensue in the coming years, with Sproul at its epicenter. But it wasn't necessarily the first claim, even that year.

The Rivals

So, what was the first purported exoplanet discovery? It was an old familiar place for planetary claims: the 70 Ophiuchi system.

A well-studied star, 70 Ophiuchi is 16.6 light years away from Earth. Two astronomers, Dirk Reuyl and Erik Holmberg, claimed, a few months before Strand published his findings, to have found a planet in the 70 Ophiuchi system, one even smaller than the candidate Strand would soon announce.[6] Their work didn't sit well with either Strand or van de Kamp. Perhaps there was a bit of a rivalry, though. Reuyl was van de Kamp's cousin, and he, too, was at the time studying the parallax of local stars—along with using the calculations of their trajectory to search for invisible companions.

The fog of war and the uncertainty of the claims perhaps overshadowed the announcements. With the world at war, some science news didn't quite seem so big. By April 1943, Strand seemed to be itching to return to the astronomical world, as he indicates in his letters to van de Kamp. Even while working at test facilities in Florida during the war, Strand had his head far above the clouds. He had even met other astronomers, like Edwin Hubble, stationed at various military outposts.

At first, Strand said, "Jupiter is getting more rivals!"[7] With a furlough coming that June, Strand was excited to check the work of Reuyl and Holmberg. But by July, his enthusiasm had waned. He wrote van de Kamp on July 4, suggesting they slow down publicity on 61 Cygni, admitting, "I have no desire so whatever to be hailed in public in company with Reuijl [*sic*] and Holmberg."[8] Van de Kamp believed that the rival astronomers' work was sloppy and error prone, and Strand had come to agree on this. Everything Reuyl and Holmberg seemed to say about 70 Ophiuchi would have driven the planet into its home star, and Strand attempted to further smooth out the calculations.

"It looks reasonable, but I suppose you will agree with me that it is only a possible interpretation of the residuals and not a final proof for the existence of that orbit," van de Kamp wrote after reading Strand's data. "It is, however, the result of a clean analysis, such as R. and H. were not able or willing to make."[9] Other correspondence, like a 1947 letter from Jan Schlit of Columbia University to van de Kamp, outright called the results spurious and even termed Holmberg's work on the star Castor flat out bad.[10]

Indeed, the original paper was so woefully vague that Henry Norris Russell—who gave us the Hertzsprung-Russell diagram showing stars by size, color, and mass on a curve—wrote a paper more or less confirming that their data was *workable.* "The only questionable point in their discussion is that no attempt was made to improve the orbital elements of the wide pair," he wrote, attempting to show the same hammering out of the data that Strand undertook, concluding, "Several more years of observation will be required for a good determination of the period—as the authors state."[11]

There were a few reasons to doubt the data, including personal rivalries and the vagueness of the planet's orbit. But another reason is that this wasn't the first time someone had claimed to find a planet around 70 Ophiuchi.

A History Lesson

Of all the searches in for planets outside our solar system, 70 Ophiuchi's was the first exoplanet "found," several times over. The first finding was in 1855, when William Stephen Jacob, of Madras Observatory in Chennai, India, noticed that the motions of the two stars in the system seemed to deviate ever so slightly. "We may suppose a third body to belong to the system, and to be opaque, and consequently invisible; such a body would, of course, disturb the regularity of the motions of the other two," he wrote.[12]

Like the astronomers at Sproul nearly 100 years later, Jacob focused much of his work on double stars (this term roughly refers to stars that appear near each other but aren't necessarily gravitationally bound, unlike binary stars) and astrometry. The 70 Ophiuchi system makes a tempting target for such studies. The stars are around 16.6 light years away and are easy to discern with the naked eye, through which they appear as one star. Viewed through a small telescope, however, the two stars separate from each other quite elegantly. But ever since the system was first characterized, the unusual orbit interactions have left astronomers wondering if a third companion wasn't hiding somewhere.

By 1895, Jacob's discovery claim was all but forgotten, having never gained wide favor. A ruffian by the name of Thomas Jefferson Jackson See, of the University of Chicago, rehashed Jacob's claim as his own, maintaining that he witnessed small derivations of the star caused by a "dark companion."[13] See's claim was disproved by Forrest Moulton of the same institution, catalyzing a bitter war of words, mostly one sided, between the two. See wrote a letter to the *Astronomical Journal* so vitriolic that the editors threatened heavy censorship of any subsequent submissions to the journal. "To abbreviate most effectively unfruitful discussion, Dr. See's remarks were transmitted to Mr. Moulton to afford him opportunity, if he desired, to reply; but he declines, on perfectly correct and dignified grounds, to do so; his essential and sufficient reason being that the statements are not in accordance with the facts," the journal editors wrote.[14]

See was known as a mercurial figure within astronomy. This reputation stretched back to his undergraduate years at the University of Missouri, where he had been a key player in the ouster of college president Samuel S. Laws after Laws replaced a favorite professor of See's with one who had questionable qualifications. See's testimony in the legislature regarding Laws also gained him political friends and influence within the state. But at the same time, he was accused of plagiarism by his fellow students, a situation that led to See being denied a prominent astronomical medal on campus.[15]

See's career would eventually taper off into bizarre incidents, including *more* accusations of plagiarism, more borderline libelous actions against fellow scientists, an attempt to publish an autobiography under false pretenses, and so on. Thomas Shirrell would later write that few astronomers "inspire a degree of rancour comparable to that evoked" by See.[16] Among the fights See tried to provoke was a rather bitter and completely one-sided libel against the works of Albert Einstein. See was well known but not especially well liked, and he was largely left in exile at Mare Island, off the coast of California, after being booted from the Naval Observatory in the aftermath of the 70 Ophiuchi affair. He spent much of this time at Mare Island making strange proclamations, and he died in 1962 with the reputation of a crank.

Strand attempted to rectify a third body in 70 Ophiuchi before ultimately concluding there was none. Planetary claims have seemingly died down in the 70 Ophiuchi system, though the possibility of a planet there hasn't been entirely ruled out.

Still, continued studies of the star led Strand to an inevitable conclusion. "No trace of the often assumed perturbations in the orbital motion was found," he wrote in a 1946 article in the *Publications of the American Astronomical Society*, using data from long exposures of the star to find that there was no "tug" on one of the stars implying the presence of a planet.[17]

With claims regarding 70 Ophiuchi yet again crumbling, 61 Cygni seemed to be the only reasonable case for a planet outside our solar system at the time.

At Sproul

Strand's work on 61 Cygni wasn't a chance discovery—it was the product of an ongoing hunt at Sproul, spearheaded by van de Kamp from the beginning of his time at Swarthmore.

Peter van de Kamp came to Swarthmore in 1937 from the University of Virginia. Born in Kamden, in the Holland region of the Netherlands, in 1901, Peter (Piet in his original Dutch) van de Kamp was the oldest of three boys. Peter's brother Jacob also eventually went into the sciences, albeit studying chemistry. His father, Lubbertus, was an administrator at a cigar factory, making just $400 a year, which led Peter's mother, Eugelina Cornelia Adriana van der Wal, to support socialist policies that would improve working conditions and increase wages for factory grunts like her husband. Although Lubbertus shared these politics, Peter told historian and astronomer David DeVorkin, "He was not a revolutionary or an activist or anything like that. One didn't do that so easily in those days."[18]

Peter excelled in math at secondary school and later said that he might have pursued it academically had he not taken up astronomy. He also studied violin for eight years and considered a career in music. His studies in physics had been lackluster, so he had narrowed down his choices to either chemistry or math for his future academic work when an astronomy class (called Cosmography) instilled a passion for the stars that he couldn't shake. This was in 1917, a time when Einstein was slowly upheaving our understanding of the universe. (Van de Kamp became quite a fan of Einstein's, writing him admiring letters and, on certain campus visits, even playing music with him.)

Van de Kamp studied at the University of Utrecht, receiving the equivalent of a master's degree in astronomy, mathematics, and physics. Utrecht was the only university van de Kamp could afford, tutoring other students for tuition money and living with family while he attended. He paid his aunt and uncle around US$17 per month as rent, obviously not inflation adjusted. His professors included Albert Nijland—an expert on variable stars—and W. J. H. Moll, who invented one of the earliest photometers (instruments that measure variations in light) used in modern astronomy.

After his studies, van de Kamp took a position at the Kapteyn Astronomical Laboratory, where he studied the movements of the Hyades cluster. The lab was named for and founded by astronomer Jacobus Cornelius Kapteyn, who studied the movement of stars in relation to the galaxy's rotation—the kind of work that laid a foundation for dark matter theory, that the universe has a great amount of unaccounted for mass, which can be inferred from the movement of galactic dust. Kapteyn's name is perhaps most associated today with Kapteyn's Star, a small, dim stellar object about 12.8 light years away that may be nearly 11 billion years old—making it one of the oldest stars in the Milky Way. (Our Sun is only 4.5 billion years old and will enter its dying stages before it can ever reach the age of Kapteyn's Star.) Kapteyn documented 454,875 stars in his initial catalog—a quite stellar number. Although Kapteyn Astronomical was not an observatory, the laboratory studied observations from other facilities. While there, van de Kamp studied with Pieter Johannes van Rhijn.

Van de Kamp was able to use the laboratory as a springboard to an appointment at the Leander McCormick Observatory in Virginia, a facility he worked closely with even after he resigned this position. In 1924, he took a one-year fellowship in California at the Lick Observatory, and in 1925 he earned a PhD in astronomy from the University of California, Berkeley, based on his studies of the *radial velocity* (spectra shifts due to movement) of faint stars. In 1926, he received another PhD, from the University of Groningen, in the Netherlands. He returned to McCormick shortly after, marrying Emma Basenau in May 1927 and becoming an assistant professor at its parent school, the University of Virginia, in 1928. In 1930, his daughter, Emma Marie, was born, but his marriage was strained, and after a two-year separation, he and Basenau divorced in 1936. Van de Kamp arrived in Swarthmore in 1937 to take over as the director of the Sproul Observatory. He personality recruited Kaj Strand, a Danish student studying in the Netherlands, because of his "beautiful double star work."[19]

Strand's version of events that brought him to Sproul differ a little. Strand claims that in 1935, he was at a formal event at the Paris Observatory, where a group of Dutch astronomers introduced him to van de Kamp. "I made the remark that I had read with very much interest about his father's work in a publication, and that I was very much impressed with what he had done, and van de Kamp said, 'Well, my father is not an astronomer,'" Strand told David DeVorkin and Steven Dick.[20] Strand said that he would often joke that this flattery—he had mistaken van de Kamp for a teenager or college student at the time, though van de Kamp was six years older than Strand—had gotten him the job at Sproul.

His initial work broadly studied the galactic motion of stars in the Milky Way, but he gradually pared down his study to those within a few dozen light years, gaining a significant understanding of our local neighborhood of stars and how their movements across the Milky Way relate to one another. Stars orbit the center of the galaxy in the same way that planets orbit a sun, but they don't all necessarily follow what seems like an orderly path, a la the near-circular orbit of planets. They also don't, broadly, move along a specific unerring plane as planets do—there are stars in all directions from us, up or down, celestial east or celestial west. The stars' positions also change and evolve over time. Right now, the closest star to us, Proxima Centauri, is 4.2 light years away; 70,000 years ago, a dim star called Scholz's Star and its massive but not bright failed star companion swept inside a light year of the Sun and disturbed the movements of the outermost comets. Since that time, the pair has crept to nearly 22 light years away.[21] Incidentally, the region these comets came from—the Oort Cloud—is named for Jan Oort, another Dutch astronomer and an alumnus of the Kapteyn lab.

Van de Kamp had plenty of extracurriculars to keep himself entertained. He was a voracious fan of Charlie Chaplin, collecting his films as well as other silent era comedies, like those of Buster Keaton. He was also an accomplished musician, a skill he picked up with the encouragement of his father.

But van de Kamp's passion was in the stars. In a 1950 letter, he would criticize Vassar College for taking astronomy out of its science curriculum. "Astronomy, by its very nature, provides an ideal introduction into science; it covers mechanics, physics, chemistry, and touches on other sciences," he wrote to Maud Makemson, chair of Vassar's

Astronomy Department, who had asked his help in defending the discipline. "A course in Astronomy unavoidably covers some history of science, and, at least in descriptive fashion, several terrestrial laboratory experiments." Perhaps most eloquently, he outlined the accessibility of the stars, writing, "The astronomical laboratory is the universe," that seemingly infinite place of stars, planets, nebulas, galaxies, black holes, big explosions, tiny but powerful neutron stars, and all matters of wonder and weirdness.[22]

He quickly undertook a mission of studying our nearby stellar neighborhood, chiefly on the hunt for stars with large proper motions and with a special affinity for smaller stars. One of his first big publications while at Swarthmore was the 1941 title *Mean Secular Parallaxes of Faint Stars*, which did not exactly roll off the tongue.

The Smallest Stars

Nearby faint stars tend to be of a class known as *red dwarfs*, or *M-type stars* (or M-dwarfs, if we're covering all our bases). They're reddish in color, and usually around 1/10 the mass of our Sun. Unlike our Sun and similar stars, which will last a few billion years before expanding outward and enveloping a few unlucky planets before collapsing into a white degenerate star, red dwarfs can last for trillions of years, chugging along, fusing hydrogen into helium.

These red dwarfs are dim, requiring at least a telescope to even spot and often a larger observatory to discern anything useful. But because of their low mass, they make an easy study. Most planets have been spotted by the *transit method*, observing a small amount of dimming when a planet passes across the surface of its star. Some telescopes can monitor several stars at a time, over-representing transit monitoring as an astronomical tool, as opposed to more complicated methods, though not every planet transits. Since red dwarfs are small stars (some have the same radius as Jupiter but far more mass), planets crossing their surface are easier to spot. Astrometrically, too, a planet around a red dwarf has a more pronounced effect than a similarly sized planet around a much larger star. If you're going to search for planets, red dwarf stars are, in many ways, an easy target.

But back in 1941, there wasn't a good way to spot something as small as a planet transiting even a small star. Another method that would later be used to find planets, based on radial velocity, was still in its infancy. Since an unseen companion tugs on its star, in astrometry, the star appears to move just a little bit on its otherwise straightforward path across the sky. Radial velocity looks instead at how the tug of a planet affects the motion (or velocity) of a star by breaking down its light and looking for either blueshift or redshift (or movements toward or away from the observer). Through this red light/blue light dance, astronomers can detect an unseen object tugging on the star, even just a little bit.

With astrometry, a big telescope can take a good picture of the star to see if it moves at all from its path around the galaxy. Radial velocity is an extension of astrometry—looking for minute deviations in the velocity of a star—but it wasn't a feasible method for planet hunting until the 1980s, when University of Toronto researcher Bruce Campbell perfected a method for breaking down star spectra by injecting toxic gases into a glass cell that filtered the light. Astrometry is efficient though admittedly hard to do from the ground. Two missions—Hipparcos and the aforementioned Gaia—have performed astrometry from space. Finding planets through astrometry is *really* hard. The amount planets move a star is tiny.

A 1940 letter from Gerard Kuiper to van de Kamp shows that Kuiper was already thinking beyond double stars and binaries and may have had planets on his mind. Today, Kuiper's name is best known by his eponymous Kuiper Belt, a group of small bodies on the outer edges on the solar system, each smaller than our Moon. If the four rocky inner planets of our solar system are the first zone, and the gas giants, ice giants, and their hundreds of combined moons are the second zone, the region just beyond Neptune (and sometimes even crossing over its turf) is the beginning of the third zone.

This third zone was virtually unknown in 1940. The only object known from 1930 until 1992 was Pluto, a tiny oddball world smaller than our Moon. Although (a good handful of) astronomers today classify it as a *dwarf planet*, which is considered planet-like, not an actual planet, Pluto was then considered simply an unusual small planet all alone out there.

But Kuiper's concern wasn't finding more of the belt. Kuiper, then at Yerkes Observatory, in the middle of Wisconsin farmland due north of Chicago, instead wanted to talk about one of van de Kamp's primary interests. "For some time I had been wanting to write you about a program for the search of distant companions to nearby stars," Kuiper wrote in his 1940 letter.[23] He outlined a vision of such a hunt: First, concentrate on *white dwarfs*, the remnants of Sunlike stars that have exhausted their fuel and become planet-sized balls of carbon, oxygen, and electron-stripped matter. Then, move along to *subdwarfs*, a category that encompasses red dwarfs and anything smaller. A red dwarf is the smallest kind of star; any smaller and the stars can no longer fuse hydrogen into helium, instead converting hydrogen into a slightly heavier isotope. These objects, called *brown dwarfs*, give off little light. They bridge the gaps between large planets, at around 13 times the mass of Jupiter, and a murky number between 70 and 80 times that mass. Such objects were unknown, but somewhat theorized, at the time.

Kuiper and van de Kamp had already collaborated on two star systems, Wolf 424 and Wolf 359, looking for unspecified anomalies.[24] Wolf 424 is a pair of red dwarfs orbiting each other 13.7 light years away from Earth. Wolf 359 is eight light years away and is perhaps most famous as the setting for an episode of *Star Trek: The Next Generation* called "The Best of Both Worlds."

But importantly, these letters with Kuiper hint at van de Kamp's interest in the second closest star system to the Sun. The triple stars Proxima Centauri and Alpha Centauri A and B are 4.2 and 4.3 light years away, respectively. But 6 light years from Earth (and 6.4 light years from the Centauri family) lies Barnard's Star. The star is faint and old, only 14 percent of the Sun's mass, and only 150 times the mass of Jupiter. It's invisible to the human eye, but a keen telescopic observer can find it moving across the night sky because it has one of the highest proper motions of any known star. Barnard's Star would become one of the most important stars to van de Kamp's career.

This search for low-mass stars and their companions was the intense focus of Sproul, whose 24-inch refracting telescope was suited to the challenge. By watching stars over several nights and taking photographic plates of the night sky, van de Kamp and his colleagues could watch these high proper motion stars move around the sky and witness any slight perturbation in their motion, on the hunt for binary stars.

Or planets.

Van de Kamp's interests seemed to include searching for planets with the same fervor as unseen stars. In a 1944 article in *Sky and Telescope*, "Stars or Planets?" van de Kamp outlines what they're looking for, objects in between the mass of Jupiter and the then-smallest-known star, Kruger 60B. Without addressing the objects by name, van de Kamp would inadvertently lump in brown dwarfs with gas giants, terming *planets* anything below 1/20 the mass of the Sun that emits little or no light.[25] The term *brown dwarf* was coined by preeminent SETI researcher Jill Tarter, but not until the 1980s, when such objects, intermediate in mass between a planet and a star, were on the cusp of being found.

"Unseen companions can be most effectively detected through the study of their gravitational effect on the motion of known stars or stellar systems," van de Kamp wrote in *Sky and Telescope*. A footnote to this passage makes an early case for the radial velocity method, calling it the *spectroscopic method* and noting its usefulness in finding small stellar companions. But astrometry, van de Kamp argued, is best suited for finding planets versus spectroscopic measurements, which are better suited for stars.

"The photographic search for unseen companions has hardly begun," van de Kamp predicted. "The next decade or so should witness considerable progress in this field." He would later take charge of such efforts alongside other Sproul researchers.

A New Friend

As Strand was leaving to serve with the US Air Force, a new researcher wrote van de Kamp a letter in search of work at Sproul. Sarah Lee Lippincott had just received her bachelor's degree from the University of

Pennsylvania in nearby Philadelphia and was eager to do one of two things: serve the war front effort back home or continue her astronomy studies as a research assistant.

Lippincott began her letter with apologies that she hadn't written sooner. "Due to war conditions it is very hard for me to decide," she wrote. "I have been waiting all this past week to hear from the defense plant before making my final decision." She then offered to stay for nine months as a researcher. "I am waiting to hear if you still want me and if so when you want me to start," she said, signing off, "Assuring you of my appreciation of your consideration of me, I remain sincerely, Sarah Lee Lippincott."[26]

Lippincott didn't stay for nine months. She stayed for 40 years and became a leading researcher in astrometry a powerful advocate for women in the sciences, encouraging others to apply to astronomy programs. In several file photos, Lippincott is the only woman in a sea of men.

But before any of that would happen, she would begin work on figuring out the schematics of Delta Equulei, a binary star system in which both stars are slightly more massive than the Sun. Lippincott and van de Kamp published a paper in 1945 outlining the orbital interactions between the two stars.[27] In the 1940s, Lippincott also studied Barnard's Star and another one nearby, Lalande 21185, alongside van de Kamp.

Van de Kamp and Lippincott would grow to have a close friendship, calling each other Pooh and Flip (sometimes spelled Flippe) in correspondence. That friendship and collaboration would become tied forever to the history of Sproul as an institution, especially once the planet-hunting era began.

1
2
3
4
5
6
7
8
9

A 1953 letter from Peter van de Kamp to Kaj Strand set the tone for much of what would happen at Sproul in the 1950s.

"We have found orbital motions for five single nearby stars," van de Kamp wrote. The first, Ross 614, was about to become a big deal for Sarah Lippincott. The next, BD +20 2465, or AD Leonis, emitted flares that would prove elusive—though promising— for the observatory's fledging program to study. Lalande 21185 would hover over the observatory for the better part of the 1950s, while Ci 2354 (or Ci 18 2354/Gliese 687) had a perturbance that seemed to cause of ongoing puzzlement. The last—Barnard's Star—would come to define Sproul for the better part of a decade, though not for another 10 years. The star, van de Kamp wrote, "shows definite deviations, which so far we have not been able to satisfactorily explain."[1]

Strand was, at this point, working at the Dearborn Observatory in Evanston, Illinois, on the campus of Northwestern University. Van de Kamp was buckled down as always at Sproul, continuing the hunt for unseen stars—and more covertly, unseen planets. Lippincott was working at the observatory by night while moonlighting (or sunlighting) as a graduate student, working toward a master's degree at Swarthmore.

These studies brought her toward Ross 614. All indications pointed to a large object disrupting the star there. By 1951, she had drawn out enough details to give a rough orbital motion. In 1955, she finally had a breakthrough that would establish her as a formidable astronomer—Ross 614B came meekly shining out of the data.

It's a star just 1/12 the mass of the Sun and 70,000 times less luminous. At that time, it was the smallest star ever discovered, and certainly the least massive. In 1953, finding such a star was a major accomplishment. We now know of stars smaller in radius (but far more massive) than Jupiter that crudely stretch the definition of what it means to be a star, barely crossing the threshold of turning hydrogen into helium.[2]

What's more, much like 61 Cygni's small companion, Ross 614B wasn't something Lippincott could *see* in the traditional sense at first. She saw how it affected the star and was able to deduce from there what luminosity it would need not to break through its parent star's light, as well as what mass it would need to give Ross 614A its wobble. Dirk Reuyl had suggested a binary star in 1936, but he merely found that it had an astrometric wobble, without being able to provide much more data. Lippincott was able to scrutinize it enough to figure out where the actual companion might be in reference to the larger star, Ross 614A. In turn, the Palomar Observatory, which housed a 200-inch telescope, was able to take her projections and actually see Ross 614B. Neither Sproul nor the McCormick Observatory where Reuyl worked could have made such a separation with their small scopes, barely more than one-tenth the size of Palomar's.

The search for planets is forever tied to the search for binary stars and objects in between. It's a matter of finding invisible forces at work around a distant star. The techniques used to find stars—which are massive—can be used to find far less massive bodies, such as planets. Finding something like Ross 614B is a good trial run for finding a large planet around a small star.

A news article at the time remarked that this was the third time an unseen companion had emerged out of the shadow of its larger sibling, after Sirius B and Procyon B were found in 1862 and 1896, respectively.[3] Both of those stars are white dwarfs, the remains of stars like the Sun after most of their fuel is spent. So Ross 614B was the first unseen companion discovered that was still a *main sequence star*, that is, one that's still fusing hydrogen into helium.

Resolving Ross 614A from Ross 614B was a challenge due to their closeness to each other. Ross 614B is about 4 AU away from Ross 614A, or the distance from the Sun to the asteroid belt in our solar system. And though it is less bright than its parent star, it's still giving off light, which can make distinguishing one from the other a more difficult task for observatories. Ross 614A and Ross 614B also aren't terribly far apart in size. Ross 614A is only 25 percent as massive as the Sun, while modern projections put Ross 614B at 11 percent of the Sun's mass (rather than the 1/12, or just over 8 percent, initially reported).

The press tends to latch onto stellar superlatives. The smallest, the biggest, the heaviest, the first. A new star, in and of itself, may not make the news, but something like the least massive star ever found is more likely to catch on—whether in 1955 or 2019. Finding the smallest star ever in 1955 was a big deal—Lippincott received dozens of speaking requests, had her discovery written up internationally, and gained a good bit of celebrity on campus as well. The Roberts Hall students at Swarthmore named her Miss Astronomy 1956—though it's unlikely that a Miss Astronomy 1955 or 1957 shared the honorific, which was probably made up affectionately on the spot.

Ross 614B didn't stay the least massive star ever found for long—in 1954, another Dutch astronomer, Willem Luyten, discovered an unseen companion to the star Luyten 726-8.[4] The parent star, Luyten 726-8A, had been unknown until 1949 because of its small size and mass. By 1956, astronomers at the Allegheny Observatory were surmising that *both* stars in the system were less massive than Ross 614B, which was still estimated at 0.08 solar mass. This mass estimate would later go up, but the Allegheny astronomers were right: Luyten 726-8A and B were both a hair less massive, at 0.1 solar mass each. The original estimate had placed them at 0.04 solar mass—too small to be considered a star today. Then vB10 overtook all of them. It was originally discovered in 1944 and was known to be the least luminous star. By 1962, when van Biesbroeck's catalog was published, it became known as the least massive as well—at 7.5 percent the mass of the Sun—until 1983. Several stars less massive have been found in the last two decades. Lippincott's pioneering work was perhaps impressive, too, for a reason that shouldn't have been as much of an outlier as it was: she was listed as a first author on the paper at a time when women's work in astronomy was unglamorous and frequently overshadowed.

Lippincott was something of an anomaly in the era, as any number of astronomy conferences bear out. There was a struggle afoot for women to have a place in astronomy, as in most disciplines. It goes back ages. Some women, like Annie Jump Cannon, carved out a name for themselves despite an inability to further their careers to the PhD level. Cannon devised the stellar classification system that organizes stars by their size and color, and she led the Harvard group known as Pickering's Women, low-paid astronomers who sorted through astronomical data on plates to place stars on that spectrum. Cannon was credited with classifying 350,000 stars. Much of their work was detailed in *The Glass Universe* by Dava Sobel. Henrietta Swan Leavitt and Williamina Fleming, who led the charge in Cepheid variable and Horsehead Nebula discoveries, respectively, were part of this group.[5]

Cecilia Payne-Gaposchkin became one of the first women to receive a PhD in astronomy when she graduated from Radcliffe College in 1925—because Harvard refused to grant women degrees in the sciences at the time. Radcliffe was the women's college affiliated with Harvard, though it is now defunct as an institution. She was one of the first people to suggest that the Sun was a miasma of hydrogen plasma, which met with hostility at the time. Once it was accepted, the concept helped redefine color in the Sun as a relation to the temperature at which the hydrogen was burning. At around the same time, Helen Hogg experienced a similar trajectory in getting her PhD, then went on to do groundbreaking work with variable stars.

Lippincott also had a few contemporaries, like Margaret Burbidge, who, despite attaining a PhD in astronomy in 1943, struggled to get telescope time because many observatories were still male-only institutions. She is known for her work on galaxy rotation rates and nuclear fusion within stars—as well as being the first woman president of the American Astronomical Society (AAS) and famously turning down the Annie Jump Cannon Award in 1972 because it was a gender-segregated award. (This time for women only.)

Vera Rubin struggled to get into graduate studies at Princeton and other institutions before finally getting a PhD from Georgetown. In the 1970s, she found the first evidence of dark matter in galaxies, a hypothesis that Fritz Zwicky had set forth a few decades earlier.

In a 1965 letter to a young woman in Utah, Lippincott outlined her take on gender in the field, saying, "I don't see why you should necessarily make any distinction between men and women" in astronomy. "I believe," she continued, "that many people are drawn to astronomy through the obvious fascination that almost all of us have with the night sky. Who can look beyond our earth without awe and wonderment?" That fascination, she wrote, leads many people to enter the

2.1

Sarah Lee Lippincott teaching

Courtesy Friends Historical Library of Swarthmore College

field, including women, many of whom, she urged, should be brought in to universities to speak on the topic and bring more gender parity to astronomy.[6]

But while her public profile placed her as one of the leading women of astronomy, entrenched sexism defined her Sproul career. In a 1960 *Philadelphia Bulletin* article, Lippincott describes to journalist Adolph Katz an encounter in which she had trouble getting through to someone she'd just met that her primary field was not, in fact, astrology. "The seeker of good tidings was amazed to discover that there was such a breed as a lady astronomer, figuring that the field of star-gazing was the province of greybeards and egg-heads," Katz wrote.[7] She had widespread support under van de Kamp as his most trusted colleague, but she was repeatedly denied tenure by Swarthmore College. In 1975, the same year the state of Pennsylvania distributed a "Who says women can't be scientists?" poster statewide featuring Lippincott, a visiting committee urged the college to grant her tenure, noting in its report that she had felt insecure in the position going decades back.

> Her name has been associated with the work at the Sproul Observatory so long that astronomers think of her immediately when the Swarthmore astronomy department is mentioned. Her length of service may have given her de facto tenure, but that is not sufficient to give her the feeling of security to which she is entitled. . . . If she were given legal tenure as a staff member, she would no longer feel, as she now seems to, that her position might be washed away if the research program were curtailed.[8]

Indeed, if van de Kamp was the primary face of Sproul, Lippincott wasn't far behind in her profile. He was also urging the Swarthmore to promote Lippincott, writing in one letter to Kaj Strand, "As you know, Sarah has worked with me for over a quarter of a century and is my right hand in the astrometric program. Her contribution to the Observatory work is extremely important and I dare say you are aware of this." He goes on to urge Strand—who once held a similar perch at van de Kamp's side—to write a tenure letter of recommendation for Lippincott: "The time has come that her value be recognized and a substantial salary increase be made."[9] He sent similar letters to astronomers Bart Bok and Laurence Fredrick, both of whom had previously collaborated with Lippincott. Three years later, like clockwork, van de Kamp had to send the same letter to Strand—no matter how hard he fought, Lippincott's position had to be continually renewed.

Van de Kamp stepped up several times to help his protégé, whom he called Flip, or Flippe. (She, in turn, called him Pooh.) In 1961, he wrote the president of Swarthmore, "Flip prefers the title Research Associate. She feels that lecturer has a part-time, or temporary meaning, which would disqualify her from becoming a full-fledged member of the faculty this coming year.

"This has become a matter of critical concern for her, for me, and for the department."[10]

Lippincott's love of the stars was lifelong. As a teenager, she had built a six-inch refractor telescope and convinced her father to remodel the attic of their home so she could turn it into a home observatory.[11] In a letter to Mildred Shapley Matthews, Lippincott called her entry into astronomy as a professional "completely unpremeditated; it simply seemed a happy combination of avocation and vocation which has given me a great deal of satisfaction and pleasure."[12] Perhaps this was why Lippincott coauthored, with Joseph Maron Joseph, the children's beginning astronomy book *Point to the Stars*, published in 1963. (It was one of two books she wrote, the other a guide to the city she had called home most of her life, *Philadelphia: The Unexpected City*.)

Sandra Faber, a preeminent astronomer who received the National Medal of Science in 2011, began her astronomy studies at Swarthmore. In an oral history with the American Institute of Physics in 1988, Faber called Lippincott an "inspiration" and an impetus to go further in the astronomical field. Faber wanted to earn a PhD in astrophysics—a level Lippincott never reached—to open her career possibilities more widely. She believed that not having a doctorate may have held Lippincott back from higher profile posts.

"I felt that if I had a master's degree, I would be limited to a small institution," Faber told Alan Lightman. "And therefore I would have limited access to tools."[13] Lippincott would eventually get an honorary doctorate from Villanova University, a school not far from

Swarthmore. Lippincott likely would have been able to get into a PhD program with her voluminous research, but that would have taken her away from her work at Swarthmore.

Lippincott's influence went beyond the professional influence and impetus. Faber admired her personally as well and was impressed by her independence. At the time Faber was at Swarthmore, Lippincott was unmarried and living alone, pursuing hobbies like photography outside of work. Faber told Lightman that Lippincott was "the first woman I had ever seen making it on her own." (In the interview, Faber also expresses admiration for van de Kamp, who she calls a friend who was "a fantastic and very positive influence on me.") Faber was perhaps one of the brightest women astronomers to come out of the department, alongside Nancy Grace Roman, whose work was instrumental in shepherding the construction of the Hubble Space Telescope. Roman's work was so influential that van de Kamp would joke that NASA stood for "Nancy and Several Associates." Astronaut Sally Ride also briefly studied Swarthmore, for three semesters, before transferring to UCLA, then Stanford.

The Space Age

In 1955, Peter van de Kamp gave a talk to the National Capital Astronomers called "Do Stars Have Planets?" Addressing in part why anyone would want to know about planets in other solar systems despite the seeming impossibility of directly seeing one, he cited the "growing interest in the possibilities of space travel." In 1955, "space travel" amounted to a few suborbital rocket launches, as the first orbital satellite, Sputnik, was still two years off from launch. But in the postwar era, it seemed *feasible* for the first time. Nazi Germany had made the first successful suborbital spaceflight in 1944, but it was far from an altruistic endeavor. The purpose, instead, was to test the V-2, which was designed as an intercontinental ballistic missile—a rocket that would go high up (in this case, 117 miles) and come crashing back down at maximum velocity to inflict vast damage. But others realized that such a technology could enable spaceflight, a dream of science fiction.

The idea of taking a rocket to reach beyond Earth goes back to at least the seventeenth century. But one of the earliest earnest attempts—beyond simply imagining a far future technology—came from Russian scientist Konstantin Tsiolkovsky, who took a rather interesting approach. He published a series of papers outlining his ideas for space travel, including rocketry, space stations, pressurization, a space elevator, and more. He then wove these concepts into children's stories, partly as a way to communicate his theories and give them more popular appeal. His work began in the 1890s and culminated in the 1903 publication of *Exploration of Outer Space by Means of Rocket Devices.*

The post–World War II era left many nations hoping for V-2-like technology—and not necessarily with altruistic intentions either. But the British Interplanetary Society had an idea in the same era to send a human into space aboard the Megaroc, a modified V-2. It called for placing a human in the rocket and sending them 300 miles into space, testing out human responses to various g-forces as well as space communication and solar studies. It even called for an EVA, or extravehicular activity, expedition in the five-some minutes the would-be spaceman would be in orbit. (As the BBC pointed out, the plan only ever considered men for the job.)[14] The project never made it past paper.

The postwar period was also a reckoning of sorts with the growing realization that space travel was actually possible and viewed by many as just over the horizon. Most science fiction films of the 1950s centered on space travel or alien creatures—*Destination Moon, The Day the Earth Stood Still, Flight to Mars, Radar Men from the Moon*—and on television, *Space Patrol* and *Tom Corbett, Space Cadet* had proved to be hits with audiences.

A series of articles in *Collier's* magazine from 1952 to 1954 brought space travel to a fever pitch in the public imagination. Wernher von Braun, who had gone from using Jewish forced labor in Nazi rocket labs to working for America's burgeoning rocketry program, even becoming one of its most public faces, was one of the architects of this new vision of space. Space stations, Mars exploration, and more were discussed in the pages of *Collier's*. The March 22, 1952, edition spells out exactly what the writers and editors were trying to say, with the headline "Man Will Conquer Space Soon!"

It was fairly public knowledge that van de Kamp was on the hunt for extrasolar planets. After all, the search was funded by the National Science Foundation, and was receiving the press rounds under Helen Hogg, who was working with the NSF at the time. In 1955, only two places were carrying out such operations publicly—Sproul and the Dearborn Observatory, spearheaded, of course, by Kaj Strand.[15] Strand left that institution for the US Naval Observatory (USNO) in 1958, so the program didn't go far—seemingly focused on confirming Strand's 61 Cygni observations. The press mentioned at the time that the discovery of Ross 614B may have enabled this hunt.[16]

But the Sproul program was a little more prodigious, with van de Kamp announcing several post-Cygni candidates. Reported planets in the systems of Kruger 60A and Eta Cassiopeiae were announced in 1956. Both were considered provisional at the time and didn't catch on widely. In 1958 two candidates had a little more longevity, BD +20 2465/AD Leonis and Ci 2354/Gliese 687. All those companions were believed to be between 10 and 20 Jupiter masses, making them enormous if they were planets. These masses—especially compared to their home stars—led exoplanet pioneer Shiv S. Kumar to debate what made up the difference between stars and planets, and what consideration we should give some of these objects, given their influence on their stars.[17] The planets didn't, according to Kumar, orbit the stars in the traditional sense. Instead, they orbited each other in a constant game of tug-of-war across a center line tilted somewhat in the star's favor. This is how the planets could be detected astrometrically. By 1967, Kumar was even supposing that perhaps these objects had formed as stars and then slowly contracted while losing mass.[18] He may not have known it at the time, but he was laying the groundwork for the debate over what is a planet and what is a stranger kind of object called a brown dwarf.

Van de Kamp's talk to the National Capital Astronomers brought up several interesting foreshadowings, as reported by Benjamin Adelman, editor of the *Junior Astronomer*. (Adelman and his brother, Saul, would go on to write *Bound for the Stars,* a book about interstellar travel.) Among them was the idea of witnessing transits, tiny drops in light when a planet passes in front of its star. Adelman wrote, "The best way to do it [witness transits] would be to watch as many stars as possible with these instruments all the time; that is, to have a continual watch by astronomers all over the globe. That would take a lot of telescopes and a lot of photo-electric photometers and a lot of astronomers. There just aren't that many astronomers and there are lots of other things to do."[19]

It took nearly 50 years, until the first decade of the twenty-first century, but astronomers somewhat did what Adelman described—they just did it mostly from space. Transit telescopes like CoRoT, SWEEPS, and MOST made it possible to detect exoplanets via transit, while the Kepler space telescope stared at the same region of sky and found planets by the thousands. Transit telescope programs were the first to make genuine concerted efforts to survey stars for transiting planets, paving the way for Kepler and TESS. Transit photometers are now sufficiently cheap and sensitive that a backyard telescope could, theoretically, find transiting exoplanets using not much more than a cheap 70-cm telescope and a photometer as an eyepiece.

Lalande 21185

The space age was off to a heated start. By 1957, the Soviets had launched Sputnik 1 into space, the first human-made object to orbit the Earth. The US Army followed suit the next year with Explorer I. This would kick off a semipeaceful brinksmanship that would see the first humans in space by 1961.

Around the same time, in 1960, Sarah Lee Lippincott culminated her work on the then-fourth closest star to Earth, Lalande 21185. It had been on Sproul's radar for a while, and Lippincott had put in a good bit of work on the star. By 1951, Lippincott and van de Kamp noted that they were seeing *something* in the Lalande 21185 data, but they didn't have too many details on what, exactly, it was. They estimated that the object could be anywhere from 3 to 11 percent the mass of the Sun and a faint companion not unlike that of Procyon or Sirius.[20]

But all they had at that time were rough estimates. In the intervening nine years, Lippincott drew out more details of the companion. The first announcement came at the May 1960 AAS meeting, where Lippincott gave provisional results. The first small publication came out in August 1960, when Lippincott reported an object about 1 percent the mass of the Sun, or 10.4 Jupiter masses.[21] The next month, she had more details—including the suggestion that the companion could be seen in infrared under the right conditions. She also noted what would become a problem for Sproul: every once in a while, their parallax measurements were off just a little bit on the *x*-axis, which had to be accounted for in their measurements.[22]

For the second time in its history, Sproul thus had a strong case for a planet, as well as four weaker cases. While Kruger 60A and Eta Cassiopeiae's planetary companions would eventually fall out of favor, BD +20 2465 and Ci 2354 stuck around. (61 Cygni also continued to be seen as a strong candidate.) It seemed that already, Sproul's planetary hunt program was yielding results.

Still, since astronomers were refining the star-planet boundary at the time, they couldn't be entirely sure if the object at Lalande 21185 was a very small star or a very large planet. An article in the *Austin Statesman* mentioned that Sproul was also trying to hammer out the details on . . . something at another nearby star, Barnard's Star, and that astronomers weren't sure what they were reckoning with there either. The same article mentioned Project Ozma (the first dedicated hunt for extraterrestrial technology) in the kicker, albeit not by name.[23] The *Baltimore Sun*, *New York Times*, and other papers picked up on the story, although not every news outlet got it quite right. The now-defunct magazine *Science and Mechanics* opened up by asking if Lalande 21185 was proof of a tenth planet in our solar system.[24]

A 1960 article from the *Philadelphia Bulletin* also tied some of the planetary research into the first dedicated searches for life, though Project Ozma aimed at Epsilon Eridani and Tau Ceti rather than any of Sproul's candidates. To be fair, though, Sproul's targets were considered quite large by planetary standards, and few researchers believed gas giants are hospitable to life, much less what could be very, very small stars. (Carl Sagan, one of the most imaginative voices in science, put forth a possible chain of bizarre life on Jupiter-like worlds, what he called "floaters, sinkers, and hunters," in his 1980 PBS series *Cosmos,* though he proposed it more as a thought experiment.)

Lippincott told the *Bulletin*, "We've been conditioned to talk about (other-worldly life) partly by the satellites and the prospect of space exploration."[25] This is part of the particular allure of finding planets outside our solar system, however inhospitable they may be—and no serious candidate appeared to be anything close to hospitable to life until 2005, more than 100 years after See's claims, and more than 60 years after 61 Cygni (a better benchmark for the beginning of the extrasolar planet hunt, with more reputable researchers at the helm).

There was plenty of activity in Sproul in the early 1960s—including van de Kamp's speculation, which today seems likely correct, that the Sun may have been formed alongside another, long-lost sibling star. "This might be discovered by astronomers someday, but it will be a tough problem," van de Kamp said.[26] There's been quite a volley over whether stars form on their own or in pairs, but a 2017 study suggested that the density of star-forming regions means that stars likely start out life in a barycentric dance with a stellar companion.[27]

But in 1963, companion stars would seem like old news after the Sproul staff made an announcement that would shake up the astronomy world.

1
2
3
4
5
6
7
8
9

In April 1963, van de Kamp unveiled his results at a meeting of the American Astronomical Society. There was a planet around Barnard's Star, the second closest star system to the Sun.

Van de Kamp's study, published in the *Astronomical Journal*, lays out the case for the now famous planet. As Barnard's Star zipped in its path across the sky, van de Kamp essentially reported that it did not stay in a straight line, but rather veered in tiny, but detectable, ways along its path—much smaller motions than a companion star would create. In fact, van de Kamp believed an object just 1.6 times the mass of Jupiter was lurching around the star every 24 years.[1]

Barnard's Star is a small, dim, old star just 6 light years away, or "36 million million miles," as one press clipping put it. It zips across the sky with the highest proper motion of any known star. If someone were to look for a planet around it through astrometry, it would make a fairly easy target. It's just 14 percent the mass of the Sun, so any disturbance in its path would be fairly detectable astrometrically.

The announcement was the culmination of 25 years of work at the observatory, stretching back to the beginning of van de Kamp's tenure at Swarthmore. The press announcement from Swarthmore places the hunt as kicking off in 1938. Indeed, a 1940 letter from van de Kamp to Gerard Kuiper requests more information on a star called BD +4 3562 and its usefulness as a reference star for Barnard's, and van de Kamp gave talks as early as 1944 about his work on Lalande 21185 and Barnard's Star.[2] Van de Kamp had previously discussed a disturbance with Strand.[3] A 1955 letter from Edwin Dennison of the Upper Air Research Observatory indicates that van de Kamp was actively studying the star at the time—though Sproul may have still thought they'd find a companion star rather than a companion planet.[4]

By 1962, Van de Kamp's hunt for a Barnard's Star exoplanet was near the level of an open secret. He frequently corresponded with John Lear, the science editor of the *Saturday Review*, hinting that something was on the horizon. "This project is being pursued in an intensive manner and I hope to have some positive results in the 'near future,'" van de Kamp wrote. "I can easily draw some hasty conclusions which might be convincing to others but not to me."[5]

Van de Kamp explained that he wanted to work on the problem until he could "remove any reasonable doubt," and promised to keep Lear in the loop on further developments.

Making the measurement necessary to demonstrate the existence of a Barnard's Star exoplanet was painstaking. The size of the deviation from a "straight" line across the sky was just 1/20,000 of an inch on the Sproul plates. (For the astronomically minded, that's a 0.0245 arc second difference.) Van de Kamp deduced from that measurement that the planet was about 1/700 the mass of the Sun, which is far below the dividing line between star and . . . not-star.[6] The discovery was at the very limits of what was possible at Sproul, but van de Kamp was able to draw it out nonetheless.

The data stretched back as far as 1916 and involved 2,413 different plates from 619 nights to compile the picture of a world that hovers between the temperatures of Saturn and Uranus, out from its star 4.4 times the distance of the Earth to the Sun, measured in astronomical units. Jupiter is 5.2 AU out in a 12-year orbit, making the Barnard's Star B orbit comparatively slower. The coldest temperature ever recorded on Earth was −128.6°F, or −89.2°C. The estimated average temperature at Barnard's Star B hovered much closer to −300°F or −184°C. By comparison, Jupiter is a balmier −234°F. That's far, far too cold for life, and the planet's high mass meant that it had to be a gas giant, so there's no solid surface to stand on. It's cold, big, and lifeless. (Modern modeling of a planet at that distance and period could estimate even colder temperatures. A planet nicknamed Hoth by NASA, but known to astronomers as OGLE-2005-BLG-390Lb, has a similar orbit and star to Barnard's Star B and is estimated to reach as low as −370°F or −220°C.) At those temperatures, van de Kamp told *Newsweek*, Barnard's Star B would be "just a horrible place for life."[7]

A New Era

At this point in Sproul history, the observatory had two planets already under its belt with 61 Cygni and Lalande 21185. Those discoveries had made ripples, but the Barnard's Star discovery was making outright waves.

NASA and the Russians were looking at leaving Earth for the first time in that period, first to the Moon and then beyond, robotically. This was making people think of where these space travel ambitions might one day land us—far beyond the solar system. Maybe the strong public interest stemmed from the creep of science fiction from pulp fodder to cinematic and small screen audiences. Maybe it was the unfortunate timing of 61 Cygni's discovery amid a worldwide conflict that shook up the entire balance of world politics. But none of this really explains the discrepancy between the huge media response to the Barnard's Star discovery and the relatively more muted press coverage of Lalande 21185 just a few years earlier.

Perhaps the disproportionate response to the Barnard's Star discovery was related to the emergence of the "flying saucer" phenomenon, which put space on the minds of the public. Van de Kamp was asked by the Air Force in the late 1950s to comment on the emerging sightings of weird lights in the sky. He told an Air Force major, "I am an astronomer, have never observed a bona fide 'flying saucer,' and have the impression that there is a strong correlation between the incidence of 'flying saucers' and the state of mind of an observer influenced by fear, hysteria, wishful thinking or plain fraudulent intentions."[8] Few astronomers took the phenomenon seriously. Pluto discoverer Clyde Tombaugh expressed a moderate interest in the 1940s and 1950s, while J. Allen Hynek briefly commandeered Dearborn to look for UFOs and spent much of his life investigating them after an initially promising astronomy career.

Outside Ufology, serious scientists were considering the possibility of alien life: Project Ozma, the first SETI experiment aimed beyond our solar system, had kicked off in 1959. Press coverage of Ozma by the *Philadelphia Bulletin* showed a hungriness and readiness—and perhaps restlessness—to find planets beyond Earth and showed that the Sproul staff wasn't alone in the hunt. Astronomers like Su-Shu Huang were already trying to figure out what kinds of planets could host life.[9]

Huang's work was the first to define and attempt to refine the idea of a habitable zone.[10] His work determined that stars slightly larger and slightly smaller than the Sun could make for the right temperatures for liquid water to persist, but this also virtually ruled out most small stars like Barnard's—dim and riddled with flare events that could cook any planet close enough to get the right heat for life. The debate about their habitability persists today, but for years, the astronomy community ruled out red dwarfs entirely.

That meant that two out of the three Sproul planets were completely out of contention. (Both stars in 61 Cygni are closer to the Sun in size.) The Barnard's Star planet was also colder and even more remote than the one around Lalande 21185. Maybe the mass alone of the Barnard's Star companion made it almost definitely a planet rather than a small or failed star. The term *brown dwarf* wasn't in use until much later, and most candidates were simply referred to as "substellar companions." Some researchers also called these low-mass objects "black dwarfs," though that name has since been applied to completely dead stars.

There wouldn't be any life at −300°F, as was estimated for the planet, and Barnard's Star B had no solid surface to stand on. Yet the discovery still sparked the public imagination. The newswire Science Service reported, "Discovery of this planet outside the sun's family means that the universe abounds with billions of other planets," and "Astronomers estimate that a hundred million of them have some form of life. Some may even have life as advanced or more advanced than human beings."[11] *Newsweek* even proclaimed, "Little by little, man's sense of his uniqueness in the universe is shrinking as his knowledge grows."[12] Science Service and the *New York Times* both gesture at the masses of Lalande 21185 B and 61 Cygni C as being between those of a planet and a star, though both were estimated to be about 10 times the mass of Jupiter—still planets by our current definition, if behemoth planets.

Keen backyard astronomers wouldn't have been able to see the planet for themselves. "Despite its proximity to earth, Barnard's Star is invisible without a telescope because it is so dim," the *Times* wrote. "Likewise, the new planet is also invisible—even with a telescope."[13] The brightness of stars and other celestial objects is measured in *magnitude*. The Sun is −27 magnitude, and a full moon is −13. The planets creep closer to a zero magnitude: Venus at −5, Jupiter and Mars at −3, and Mercury at −2. Hands down, the brightest star outside the Sun is Sirius A, which shines at a magnitude of −1.46. Saturn and the nearby very, very bright star Vega are magnitude 0.

From there, as you count up, stars and planets get dimmer and dimmer. Uranus, which is mostly drowned out by city lights, is magnitude 4, and Neptune, which can barely be seen by the naked eye on a good night in a dark field, is magnitude 8. Ratcheting that all the way forward, Barnard's Star is a 9.5 magnitude star, which requires a telescope to spot. Hubble can go up to 31 magnitude, where it can capture galaxies from near the beginning of time. Yet it can spot only very hot, very new planets because of how dim these bodies are compared to their star. Planets, after all, don't produce light—they only reflect it and give off heat.

At the time, Barnard's Star B was believed to be 30 magnitude.[14] This was far outside the reach of even the most powerful telescope of the time, the Hale telescope at Palomar Observatory, in California. Where the Sproul refractor was 24 inches, Hale's was 200—and its initial test run in 1949 pushed it to only 20 magnitude, with a great amount of time used to expose the stars photographically.[15] Thus, even a 200-inch, 14.5 ton mirror couldn't find a planet directly in 1963.

This is a long way of saying that finding a planet is really hard. The slightest shift in a star's motion gives away the presence of gas giants in the system. Peter van de Kamp looked for deviations on a straight line, or "wobbles." In finding these wobbles, he revealed a way to detect planets without seeing them, using a method normally employed only for stars. The technique opened up a new world to astronomers, moving the discipline toward being able to discover other worlds. A cold, inert gaseous mass of a planet seemed to herald the age of planetary discovery.

Van de Kamp's Rise

The discovery boosted van de Kamp's public profile and fueled interest in planets outside our solar system. Just three years later, in 1966, *Star Trek* debuted. The American Astronomical Society put van de Kamp on its national committee for the wider International Astronomers Union in late summer 1964, presiding over a commission on double stars and giving lectures on extrasolar planets. In the September 1966 issue of *Esquire*, he was drawn in the style of a Marvel Comics character, listed as a "super-prof." But in his own style, van de Kamp told the campus paper that his superpowered persona was perhaps superfluous. "Marvel Comics? Charlie Chaplin is good enough for me," he told the paper, as well as expressing gratitude that the magazine had spelled his name correctly.[16]

His research was covered in *We Are Not Alone*, a book by Walter Sullivan often mistakenly shelved away in the UFO/Paranormal sections of used bookstores, maybe because of its superficial similarity to less reputable books of the era like *The Flying Saucers Are Real!*, *Strangers from the Sky*, *From Outer Space to You*, and so many other cheapie paperbacks. But Sullivan wasn't a quack—he was a *New York Times* reporter. And the book wasn't about UFOs, which don't warrant a mention. Instead, it opens with a meeting of the secretive Order of the Dolphin, which wasn't an illuminati conspiracy so much as a gathering of scientists and engineers to discuss the nascent field implied by the public name the group decided on: Search for Extraterrestrial Intelligence, or SETI.

The book, like many newspaper articles of the era, ties SETI indirectly to Sproul's work. This makes perfect sense—if you're going to find alien life, it needs to come from somewhere, and in 1964, when the book was written, only a scant handful of data points existed regarding planets outside our solar system. The book also points out an important role 61 Cygni served, aside from home of the first purported exoplanet: Friedrich Wilhelm Bessel used the apparent motions of 61 Cygni to determine the path Earth takes around the Sun, then calculated the distance to the star using simple trigonometry.[17] (Bessel was about a light year off, but his estimates didn't have the benefit of Hipparcos or Gaia data.)

We Are Not Alone discusses Strand and van de Kamp's work on the star, and Lippincott's work on Lalande 21185, before mentioning the Barnard's Star discovery. Van de Kamp had told Sullivan that the apparent object "must definitely be regarded as a planet and can shine only by reflected light."[18] Sullivan sought to define the boundary between two kinds of objects: objects neither stars nor planets, but something in between. We call them brown dwarfs today, and the boundary between them is . . . fuzzy. Sullivan called L 726-8 the smallest boundary at which something still shines by its own visible light, and anything below that a . . . not-star. "An object as large as the invisible companion of 61 Cygni may glow a deep red because of its heat, but it is not a ball of incandescent gas like the Sun," he wrote. "For every star intrinsically brighter than the sun, in our neighborhood, we see at least twenty fainter stars. If this pattern continues into even smaller sizes, they are very numerous indeed."[19]

In 1966, van de Kamp appeared in a PBS special hosted by Don "Mr. Wizard" Herbert—"The Invisible Planet," one of an eight-part series called *Experiment: The Story of a Scientific Search*. The special laid out the painstaking work performed by the Sproul staff. Herbert explained that "If the star should have an unseen companion, the amount of wobble or perturbation from a straight line, measured on the plates, could be so small as to approach the lowest point to which errors in the technique can be reduced. The astronomer constantly has to ask himself whether the star is really wobbling or only seems to be wobbling because of undetected errors in the observations and measurements."[20]

The program opens with van de Kamp playing piano for Olga van de Kamp, his wife, before he trots off through campus, upstairs, and into the observatory, with a big, dramatic reveal for the telescope. It then pans to the interview with Herbert. Far into the program, he asks van de Kamp, "It looks like Barnard's Star is wobbling across the sky. Could its wobble be caused by a body orbiting around it that's invisible? Or by a hidden error—an undetected factor somewhere in their [astronomers'] measurements that could produce a false wobble?" Van de Kamp then speaks of the rigor of the staff, confirming that the instrument is correctly calibrated, and that they're not seeing something that isn't there—especially with such a small margin of error in the telescope. "Finally by 1963, I decided, well, I guess we better make some statement," van de Kamp says. "We had over 25 years of observations and dates—there was no escaping the conclusion."

The program also gives a stark demonstration to explain the amount of wobble. Herbert pulls out a piece of tissue paper, which he measures at one-thousandth of an inch. In that small amount of space, Herbert says, the planet could move around Barnard's Star from Sproul's perspective with plenty of room to spare. (Herbert also states that Barnard's Star holds the crown as the first definitive planet found, though the unseen objects of 61 Cygni and Lalande 21185 were still considered planets by many.)

Van de Kamp was called on frequently to give talks and interviews in the wake of the Barnard's Star discovery, his name appearing everywhere from a philosophy conference to articles about human spaceflight and UFOs. At the conference, he laid out his guiding principles, saying that science can only survive the increasing influence of the military-industrial complex and Cold War paranoia "if we want to do science for its own sake."

"It is unfortunate that the achievements of both science and art can be used for baser purposes," van de Kamp said. "At this festive occasion I do not wish to dwell upon the wholesale harm which is done to humanity by applying science, and as far as that goes, the art of demagoguery for destructive purposes. All of us are living under the shadow of potential annihilation, and are desperately in need of human wisdom, which is often so woefully lacking."[21] This was tied to his criticism of the nascent human space exploration program. He told the *Philadelphia Inquirer*, "It is motivated by the military and politicians," and remarked that the idea of space astronomy was a "provincial approach," and that there was "nothing to replace a large surface telescope and surface astronomy with the earth as a platform. . . . Besides, you can breathe."[22] He also found lunar ambitions a little silly, saying, "Before 1958, no astronomer would be caught dead studying the moon. I mean, it just wasn't the thing to do."[23] He found that efforts to get to the Moon were, above all, based in military ambitions:

> The billions of dollars that have been spent on space ventures have led to some interesting results, but, as I said this afternoon, this money would never have been spent for pure science; it is simply being spent for military and political purposes.
>
> As a by-product there are some rather interesting scientific results, although these by themselves would not have justified the effort. One never would have gotten the money for it, because you can't convince people to pay money for things like that; you can scare them into paying money for defense.[24]

These criticisms go as far back as the beginning of the space race. In a speech revolving around the Soviet space dog Laika, who yelped her way to a fiery end aboard a Sputnik capsule, van de Kamp said:

> Sputnik has brought out the worst and the best in us. Radio and newspapers again blare forth the stereotyped, "We must beat the Russians. We now have, or will have, missiles that can hit Moscow," thus proclaiming unconvincingly our peaceful intentions to the world, which by now is a bit concerned about American ideals and policy. How would we have reacted if the English or the Danes had been first?—Do these harbingers of fear forget the peaceful intentions of the International Geophysical Year of which the Sputnik launchings are a part[?] . . .
>
> As to the significance of sputniks, various aspects need discussion. Sputniks represent a formidable achievement, but not a scientific breakthrough. Obviously—as any astronomer can calculate—a substantial amount of energy and know-how is required, first, to put a sputnik [*sic*] several hundred miles above the earth's surface; next, to direct it in a near-circular orbit. This aiming of the Sputnik is not only a matter of brute force, but of very careful scientific adjustment. . . .
>
> Probably, Sputnik has its implications in a struggle or race for military supremacy. However, I would have to be shown that Sputnik and its fellow travelers, the rockets and missiles, can either win or deter a war. Sputniks themselves have little or no military value, except that they indicate the existence of intercontinental ballistic missiles.[25]

In this speech, he inadvertently foreshadowed efforts that would come in the wake of his discovery of Barnard's Star B, beginning with chatter soon after of how we could get there. But in 1957, van de Kamp was saying, "We had better not talk about man's space travel towards the stars. Even traveling at as high a speed as 50,000 miles per hour, in other words, three times the speed of the present sputniks, it would take 80,000 years to get to the nearest star, or over 2,000 generations. Enough said!"[26]

But his speech also foreshadowed one of the biggest problems of the 1960s as well: military engagement. While van de Kamp was ramping up his research, the Vietnam War was heating up, causing strife and discontent across the United States. Students—including those at Swarthmore—were immune to neither the unrest nor the prospect of the draft. Soon, the Swarthmore campus would become a site of protests against the war and the ever-burgeoning militarism of the Cold War era.

Meanwhile, Kaj Strand saw to the completion of a 61-inch telescope at the USNO station in Flagstaff, Arizona. At 2.5 times the size of the Sproul refractor, its astrometric program could have made the work of studying nearby stars much easier. It wasn't used much for planet hunting in the end, though, but it did discover a new world in 1978: Charon, Pluto's largest moon, was found using this telescope, which as since been named after Strand.

On Campus

At Swarthmore, a big surge in astronomical interest followed van de Kamp's discovery of Barnard's Star B. His fall enrollment jumped from 60 to 90 students after the announcement—up from the dozen or so students in 1943, around the time of the 61 Cygni planet discovery.[27] The Sproul Observatory was expanded, which was a boon to an observatory that once had wings literally used more or less as a root cellar, according to Nancy Grace Roman. She studied astronomy at the time of the 61 Cygni affair, and said of the observatory, "To give you an indication of the condition of the observatory when we started, it was being used for storing onions! And that was really a pretty good clue to the condition of things in general." (This wing, which contained a six-inch and a nine-inch telescope, was typically used by astronomy students.) Roman also recalled low enrollment in classes during the war, including a course where it was just her and van de Kamp, who still managed to give two-hour-plus lectures to his audience of one.[28] Roman notes that some of this had to do with so many people of college age being enlisted, leaving a kind of low-morale ghost town on campus.

The discovery of the Barnard's Star planet was also warmly received on campus. Daniel Hoffman, a professor at Swarthmore, included it in a poem written for the campus's Phi Beta Kappa Honor Society:

> **Imperfect learning, bless this place**
> **With possibilities of grace.**
> **Let Mind, that ranges Heaven as far**
> **As Barnard's lightless, pendant star,**
> **Discern, though darkness shroud the soul,**
> **Its constant, living aureole**
> **That casts one comprehending light**
> **Across our chaos and the night.**[29]

Hoffman would go on to be the consultant in poetry at the Library of Congress in 1973 and 1974, a position we know today as the poet laureate.

At the beginning of 1965, the NSF gave Sproul $49,900 to continue its research into low-mass companions. The money was to go specifically toward studying stars within 16 light years of Earth in the search for either small stars or planets over the course of two years.[30] Van de Kamp also received an "endowed professorship" that gave the Astronomy Department an additional $400,000 boost.[31]

It didn't take long to yield results. In 1966, Sproul announced a new planet, this time around the star BD+5 1668, also known as Luyten's Star. The Earth-mass, possibly habitable planet is today known as the target of a music-filled message beamed by a group called Messages to Extraterrestrial Intelligence on the off-chance aliens were listening in. It's about 12 light years away. In 1948, van de Kamp published a study on the star, noting a slight shift every six years, but said that the movements were so small that it could just be an instrument error.[32] Willem Jacob Luyten, for whom the star is named, was a fellow

3.1

Peter van de Kamp
Courtesy Friends Historical Library of Swarthmore College

Dutchman and colleague of van de Kamp and a visiting professor at Swarthmore in 1969.

But—perhaps emboldened by the Barnard's Star discovery—van de Kamp took a similar look at Luyten's Star and determined that a planet of two Jupiter masses orbited the star. "The possible unseen companion was detected only by pushing measurements to their limit," Ann Ewing, of *Science News*, wrote. "Astronomers hope future observations will confirm the tentative identification."[33]

Confirmation is a big hurdle in planetary finding. As of January 2018, NASA's venerable Kepler spacecraft, specifically designed for planet hunting, has found 2,341 planets—but has another 4,496 candidates awaiting confirmation. The second mission, K2, has 185 confirmed planets but 515 candidates waiting in the wings, meaning that *more than 5,000* candidate planets await confirmation.[34] The follow-ups will come primarily from ground-based observatories, with Hubble, Spitzer, and, sometime in the 2020s, the James Webb Space Telescope making additional observations of especially juicy targets. Confirmation may require a bigger telescope to patiently wait a few transits; a telescope using adaptive optics to look for a hint of the planet; or use of the radial velocity method, that other wobble detection method. False detections and activity intrinsic to a star (like star spots) also need to be accounted for.

But not all 5,000-plus candidates will wind up being planets. Like Sproul with Luyten's Star, the Kepler candidates pushed Kepler to the limits of what would or wouldn't qualify as a planetary transit. The little dips in light associated with a planet could be nothing at all, just an instrument error or something else flying between Kepler and the star at the right moment.

Thus, the Luyten's Star discovery seemed to make less of a splash than any of its three Sproul "siblings," being quietly published in science magazines but failing to make much of an impression in the wider press. Little was known about this world, and it was presented at a conference proceeding but not published in a journal. It was, however, included in Sproul literature on the subject of unseen companions, establishing it as a pretty firm candidate.[35]

Changing Times

In 1968, John Hershey came to Swarthmore as an assistant professor in astronomy. He dove into the work on nearby stars, with his early work especially focused on CC 1213, which van de Kamp described as showing a "nice perturbation" in a letter to Kaj Strand, though CC 1213 didn't seem to pan out in the end.[36]

Hershey came at a time when Swarthmore—and the nation—was in great tumult. There was a strange contrast between the concerns of the interstellar—where Earth is but a small island in an unimaginably massive sea—and the real, visceral, personal, political turmoil at home. Swarthmore wasn't immune from student protest. The school was Quaker chartered, leaving it hard to put down on a Right-Left spectrum—pacifism and relatively egalitarian tendencies existed alongside more modest and traditional trappings of personal behavior. Thus, while it might have been an integrated school, that did not necessarily make it liberal to the likings of the 1960s New Left. Sex was prohibited, for example, and students with beards weren't allowed to be tour guides for the campus.[37]

Swarthmore president Courtney C. Smith sat at this odd juncture. Smith, who became president in 1953 at the age of 36, had made national news by speaking out against McCarthyism and leading a protest against an academic loyalty oath. He also stood up for students bringing politically radical speakers in, such as Gus Hall, the leader of Communist Party USA. Prior to his academic appointments, he had fought to upgrade facilities for black cadets at Naval training grounds in Pensacola, Florida. He was a committed environmentalist and, possibly as an outgrowth of Quaker pacifism, worried about the spread of the military-industrial complex onto college campuses.

But it would be wrong to call Smith a *radical*. While he believed that a campus was a "matrix of social justice," he did not see it as a "direct instrument" for the enacting such justice.[38] In a letter to van de Kamp in 1964, James Wanner, a PhD student at Harvard who had interviewed for a position at Swarthmore, described having made a slight flub in judgment. After hearing that the college had a more laid-back approach to attire, he had arrived at his interview in an ascot (rather than a more businesslike necktie). Wanner's information had come from a student who had judged that "his professors were non-conformist and informal in their classroom dress."

"In talking to Dr. Liller yesterday I was surprised and sorry to learn that my preference for ascots seems to have played so big a part in the current hassle," Wanner wrote van de Kamp. He wondered "if and when President Smith is assured of [his] maturity" whether he would be considered for a teaching position. (Van de Kamp's response was not in the Swarthmore archives.)[39]

Life magazine described the Swarthmore president's complexity: "Dr. Smith's personal principles combined an old-fashioned morality with a lofty liberal ethic that made him contest McCarthyite assaults on academic freedom in the '50s when most college presidents were maintaining a timid silence." It was under Smith that the campus also became openly integrationist.[40]

But student tensions ramped up throughout the 1960s, far, far beyond the concerns of ascots and into revolutionary, antiwar fervor—which Smith was unprepared for. After weeks of intensive student protests over racial tensions on campus, Smith died in his office, drawing a close to the student protest on a bitter note, which further drove tensions on campus.

Black students were a small fraction of the student body at Swarthmore. At the time, they represented 47 out of 1,024 students on campus. So the Swarthmore Afro-American Student Society (SASS) was formed to organize for better representation. The group's members felt that they weren't given much of a voice on campus, and they argued that a report by dean of admissions Lawrence Hargedon had recently outed them based on their backgrounds. Several of the students felt their privacy had been violated. They demanded a black assistant dean of admissions and a hand in guiding how the campus recruited minority students—or at least the right not to have the decisions made by well-intentioned but off-the-mark white administrators. When these demands went unmet, the students occupied the admissions office in a protest sit-in. The sit-in lasted a little more than a week before Smith's death led to a "moratorium." In a statement, the students said, "We sincerely believe that the death of any human being, whether he be the good President of a college or a black person trapped in our country's ghettos, is a tragedy."[41]

At the time of Smith's death, he had been negotiating with the students, promising them amnesty if the protest escalated no further, and attempting to create an ad hoc committee that included one of the small handful of black professors on campus. Part of Smith's hesitation to address the protests came from how it conflicted with his interpretation of Quaker beliefs. "We have lost something precious at Swarthmore—the feeling that force and disruptiveness are just not our way," he said. "But maybe we can see to it that this one time is only the exception that proves the rule." Beyond that, he thought, protestors were undermining "faith in education and the educational process."[42]

Paul Good wrote in *Life*, "The tragedy of Courtney Smith is a peculiarly American tragedy, devoid of villains, full of good intentions, ultimately disastrous. Perhaps it is *the* American tragedy."[43] And it's true—the black students didn't consider Smith a villain or an adversary. But they believed, "The condition of black student life in a college like this is something white liberals can't grasp," according to Clinton Etheridge, who led the SASS protest. "They automatically think they understand the racial scene, but they don't."

"Look, you can always get a good rap here, the kids are bright, and we have a commitment to that atmosphere," he said. "But there are a lot of agile black minds going to waste out there for lack of a little college support. Courtney was hung up on academic excellence. But we need a redefinition of it.

"We need some realistic humanism here."[44]

It was widely believed on campus that the stress of the situation had led to Smith's heart attack. Van de Kamp purportedly exclaimed, "they killed him," repeatedly, and later addressed a Chaplin seminar by saying, "Any student, or group of students, or any faculty member, or others who issue demands, or carry out, support, or condone any action which interferes with the functioning of the College do not belong here. They were admitted to, or joined, Swarthmore, but obviously do not understand the spirit and meaning of Swarthmore."[45]

In the wake of Smith's death, van de Kamp canceled a screening of *The Great Dictator* and showed *The Gold Rush* instead, deeming the former inappropriate in light of campus unrest. For this—which reflected attitudes similar to Smith's—he was pilloried as a conservative in the face of social change. "This seems sad and slightly ridiculous in view of my liberal, democratic and socialistic upbringing," he wrote. Indeed, van de Kamp was critical of the encroachment of military interests in campus life and matters of science, and he had held court with black students at the University of Virginia in the 1950s, when segregation publicly reigned in the south. Like Smith, he wanted a more "measured" approach, one that flew in the face of an era of direct action and direct democracy—a scenario that was playing out nationwide. "Let us face the world and our problems in a spirit which is independent and fearless," van de Kamp said. "We must think for ourselves and be responsible for our attitudes and actions. It is good and exhilarating to commit oneself to one's convictions, one of which is the dead serious obligation to fulfill our academic commitments on the highest possible level."[46]

This conflict of viewpoints pitted a perceived purity of the academy and learning versus students who viewed such ideas as a lofty outcropping of white privilege. It also soured the mood at Swarthmore, making it a place of national interest and scrutiny. The *New York Times* even excoriated the student protestors in a staff editorial:

> The aimless anti-establishmentarianism now rampant among radical students and their faculty camp-followers is sapping the strength of academic leadership itself. The university presidency, at a time of violent social tensions, is at best a storm center. But the pressures become virtually intolerable when college heads find themselves confronted with non-negotiable demands in what can only be described as a subversion of reason. However idealistic the student goals, the way they have chosen is that of the closed, totalitarian society.

> The death of Dr. Courtney C. Smith at Swarthmore College, in the face of disruptive action by a small group clamoring for more black power, appallingly underscores the price extorted by these policies of excess. . . .
>
> A few days before his death Dr. Smith said: "We have lost something precious at Swarthmore—the feeling that force and disruptiveness are just not our way." It cannot be the way of the academic community anywhere, if that community is to remain free.[47]

A few years after this incident, van de Kamp—the low-key Dutch socialist whose views had been pushed to their limit by more radical student protestors—decided to retire. But somehow, his retirement years would prove a tumultuous professional time as well, for completely different reasons.

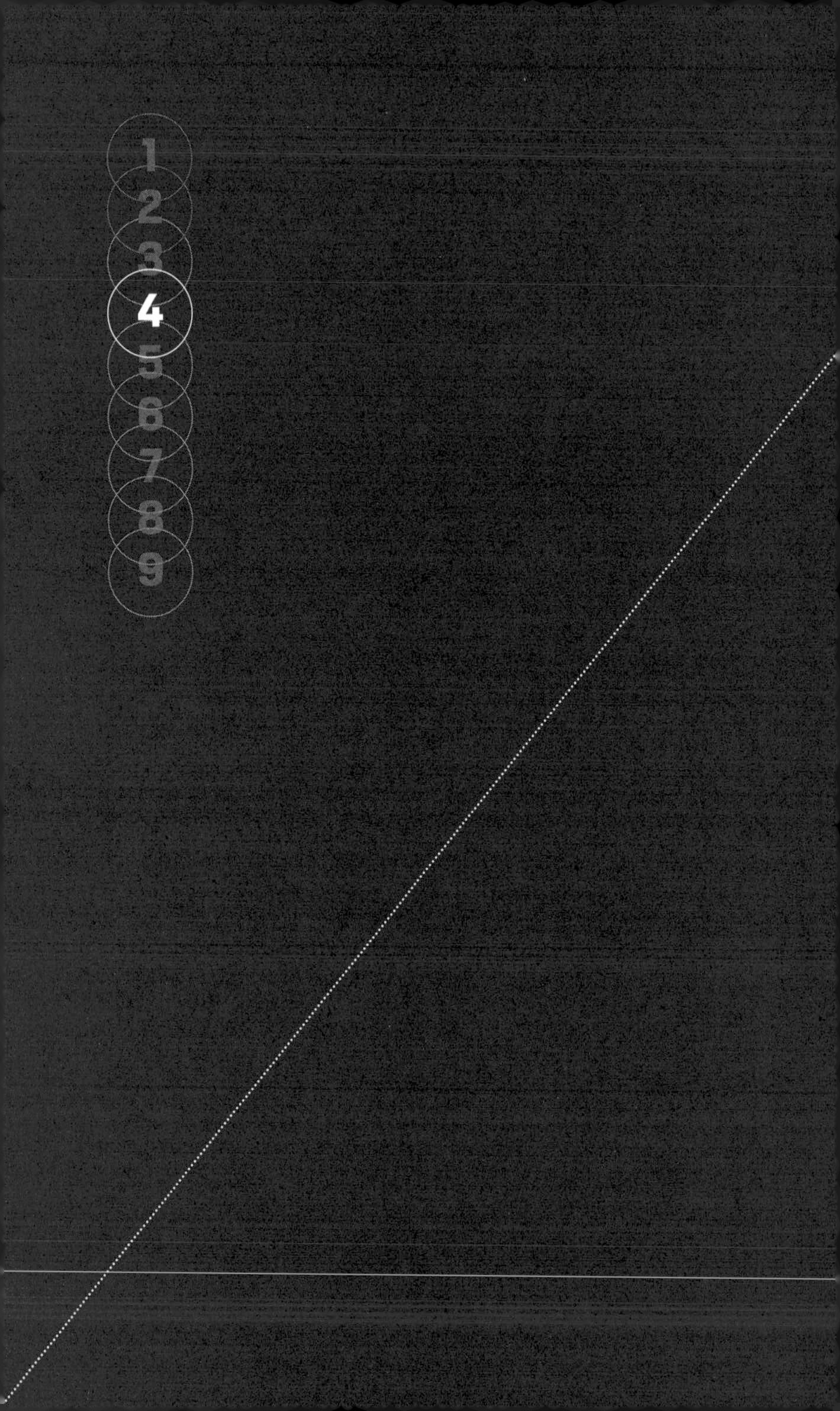

Van de Kamp began winding down his 30-plus years with the college and preparing for retirement in the decade after that, which bolstered his case for Lippincott's tenure. He was also in the process of bringing of Wulff-Dieter Heintz to the observatory from Technological University Munich.

Where van de Kamp was jovial, Heintz was deadly serious. Rather than a gregarious lecturer, he was a stern scientist, dedicated to meticulous observation of the stars, and a demeanor one colleague called "punctilious," which means "showing great attention to detail or correct behavior." That colleague, Oleksa Bilaniuk, said that Heintz's demeanor hid a wealth of subtle, dry humor.[1] He was a chain-smoking German, whose house had been bombed and his town ravaged during World War II. When Allied forces came through, he volunteered to translate for them. He was also a master chess player, who wrote multiple books on the subject.[2]

Like van de Kamp, though, he was dedicated to the study of astrometry and binary star mechanics, making him a fit for the Sproul search for unseen companions. He also had a love for educating and sharing his knowledge to spur public interest in astronomy. A visiting professorship had turned into a faculty job by 1969, with Heintz being groomed as a successor in the Sproul director role.

This friendly gamble on Heintz's future would later give way to a bitter enmity.

Van de Kamp still had some surprises in store. The mystery of Barnard's Star was deepening the more the Sproul team researched it. The proposed orbit of the planet didn't quite make sense, and there appeared to be some kind of secondary tug on Barnard's Star. Van de Kamp proposed an explanation: not one, but two planets orbited Barnard's Star, one 110 percent the mass of Jupiter and the other just 80 percent. The smaller world orbited in 12 years, the larger in 26, and both were arranged on the same plane. Though a Jupiter and Saturn comparison might spring to mind, Saturn is only 1/3 the mass of Jupiter. But the newly christened Barnard's Star B2 had a similar period to our splendid ringed world. The new measurements placed both planets in roughly circular orbits, much more like our solar system.[3]

This made the realm of Barnard's Star a bona fide planetary system and opened up the possibility that it was quite a bit like our own solar system. Under the old parameters, any Earth-size world would have been flung out unless it had been deathly close to Barnard's Star, according to Shiv S. Kumar. In a letter to van de Kamp, he wrote that "an earth-like object would not survive for more than a few hundred million years." Barnard's Star was already likely twice or more the age of the Sun.[4] The press, once again, had a field day with the discovery. Faber hadn't heard the news until reading it in *Time*, reaching out to her old professor with well wishes. "You've shown that years of cautious, careful work can yield results of unimagined precision, and very exciting results, indeed." (She also mentioned the *Life* article on President Smith and was curious to know if he found it too sympathetic to student protestors.)[5] In his chats with the press, van de Kamp called the two-planet hypothesis an "acceptable explanation" or "interpretation" of the data over a "discovery."[6]

Thus, in 1970, there were a half dozen good planetary candidates—61 Cygni, Lalande 21185, the Barnard's Star planets, the lonely gas giant of Luyten's Star, and a bizarre planetary claim not coming out of Sproul. Radio telescopes in Britain were detecting a faint murmur in the mostly regular radio beat of the Crab Nebula pulsar. Pulsars are the time keepers of the universe, emitting beams in radio frequencies with the same timing, like a violent cosmic metronome. But every 77 days, the Crab Nebula pulsar skipped a beat. It was suggested that planets could be the culprit, but that claim gradually faded away in favor of interactions with the surrounding nebula dust.[7]

The two-planet hypothesis was a capstone to his career. By 1971, van de Kamp had tied up loose ends, submitting an NSF proposal that also served as a retirement letter to his funders. In his notes to Strand, van de Kamp wrote, "Perhaps the most important aspect of my request is the need for recommending Heintz for tenure. It is hard to imagine the successful continuation of our astrometric program without the enthusiastic and knowledgeable support he is giving."[8]

Lippincott and Hershey were also recommended for tenure, which in van de Kamp's view left the observatory in good hands as he transitioned to a part-time basis by fall 1972.

Heintz was named chairman of the Astronomy Department. Lippincott was named director of the Sproul Observatory. Sproul's transition had just begun—and bumpy times were ahead.

The Star-Planet World

In early 1973, it was obvious that van de Kamp's work wasn't quite done, even in retirement. At the January AAS meeting that year, he announced a new discovery: a *star-planet* orbiting Epsilon Eridani. About 11 light years away, Epsilon Eridani is only a few hairs smaller than the Sun, making it an exciting place to hunt for Earth-like worlds. Unlike the red dwarfs of the universe, Sunlike stars are more likely to have habitable zone planets that make full rotations, rather than keeping one face toward the star at all times, like the known "habitable" red dwarf planets today. But van de Kamp wasn't claiming a habitable planet—possibly not even a planet at all. He was announcing the tentative detection of something different in a 25-year orbit around the star.

"What we have is not a star in the ordinary sense, nor is it a conventional planet," he said.[9] By the time the results were published, one estimate placed the body at 6.2 Jupiter mass, which still today would be considered a planet—though a very, very large one.[10] Using different parameters, however, the mass *could* be 52 Jupiters, which is distinctly . . . not a planet, though that is too small to fully ignite into a star.

Even today, astronomers can be hesitant to call something a planet unless they're certain it is one. Often, a planet-like body turns out be a *substellar companion* or, in certain cases, a *planetary mass object*. There's been some debate about what is a planet and what is a failed star—and what is not a failed star but just one of extraordinarily small mass. Roughly speaking, something between 13 and 70-ish Jupiter masses is called a brown dwarf, a term used to denote objects that formed like stars but never got big enough to fuse hydrogen into helium. Of course, some objects have thrown into doubt the tidy explanations of how a star forms versus how a planet forms, such as "planets" that seem to have formed on their own, independent of a star. Given their mass, they'd be something like a very small brown dwarf.

But that term wasn't in wide use until the 1970s, shortly after the announcement of the star-planet at Epsilon Eridani. Jill Tarter, who would spend her professional life largely focused on SETI, began her career on the hunt for the elusive *missing mass* of the universe, which we now call dark matter. One of Tarter's ideas was that several millions or billions of objects per galaxy, each with a mass high enough to create significant gravitational force but low enough not to ignite like a star, were all but invisible to most telescopes of the era, and are still fairly hard to find today. Tarter arrived at the term *brown dwarf* for these "almost stars" by 1975, and believed that, in concert with low-mass black holes, they could account for the vast amount of dark matter we haven't seen. (This explanation has since become known as the massive compact halo object, or MACHO, theory of dark matter, named for high-mass invisible objects.) She placed the mass at which a star wouldn't ignite at 83 Jupiter masses, but we've since found stars that dip into the low-70 Jupiter-mass regime.[11] She wasn't the first to suggest such an object existed—they were captured under the all-encompassing term *substellar*, or labeled *black dwarfs*, and astronomer Shiv Kumar theorized about them in depth—but she was the first to coin the term *brown dwarf*. She also correctly determined that, despite their mass, brown dwarfs are quite small—contracting into the size of a gas giant. Tarter didn't work much on brown dwarfs past giving them the name, as she moved along into alien hunting and providing the basis for character Eleanor Arroway in Carl Sagan's novel *Contact*.

Tarter's term *brown dwarf* obviously gained much more currency than *star-planet*, perhaps because of the cumbersome connotations of van de Kamp's term. (Other terms that have been considered but rejected over time include Lilliputian stars, infrared dwarfs, super-Jupiters, extreme red dwarfs, and substellar objects.)[12] But there had been some suggestion that gas giants were not planets themselves but very small failed stars. "Jupiter in a sense is a small star rather than a very large planet as formerly thought," Gerard Kuiper proclaimed in 1965, kicking off a firestorm of controversy. Infrared measurements of Jupiter had revealed that it was warmer than it should have been accounting only for starlight as a heating mechanism. At the time, two stars were known to be below the accepted "star" mass limit—the binary stars at L 726-8. But that conclusion had rested on what was later discovered to be a miscalculation of mass. Nonetheless, it still led astronomer Harlow Shapely to work out the idea of *crusted stars*—objects smaller than stars but larger than Jupiter that produce their own heat. Shapely also supposed this internal heating could leave habitable terrains below.[13] A better understanding of Jupiter's interior has led us to recognize that pressures are too great at those depths to sustain much.

Not Going Quietly

By March 1974, Sproul had racked up a spectacular achievement—it had taken its 100,000th plate from the intrepid 24-inch refractor, providing an intimidating portfolio of the night sky from several decades of work.[14] Those observations had yielded countless studies of the night sky and revealed some of the finest observations of nearly or completely invisible objects. Ninety thousand of those observations came from roughly 200 nights per year across 37 years, while the other 10,000 came from the first 25 years of the telescope's operation. The 100,000th plate was snapped by Michael D. Worth just as Luyten's Star crossed into the field of view. Worth was a researcher at Swarthmore who had, in van de Kamp's last year as a full-time faculty member, collaborated with him on a study of the star BD +43 4305. That star, sometimes called EV Lacertae, was purported by van de Kamp and Worth to have an object in orbit around it with a mass between 10 and 31 Jupiter masses, which would make it almost certainly a *star-planet*, to use van de Kamp's term.[15] Van de Kamp would refine that mass in later years down to 3–5 Jupiter masses.[16]

But while the retired professor had professional successes, all would not be well in 1974—that November, van de Kamp's second wife, Olga van de Kamp, passed away at the age of 72. While her husband had kept his head in the sky, Olga had been a youth psychologist who had moved to the United States after resisting the Nazis just as they came to power. At the time of her death, Peter van de Kamp was a visiting professor at the University of Amsterdam for one semester.

With van de Kamp away from Swarthmore at the time, Lippincott had to be the bearer of bad news, telling Edward Dennison that Olga had been a "dear friend" who had been confined to a wheelchair much of the past year.[17] The wheelchair had become a necessity after a decade of debilitating pain, though letters and obituaries don't specify the illness she suffered from.

Van de Kamp kept his nose to the grindstone, though, trudging through his grief with harried astronomical work and a new round of Chaplin seminars in early 1975. "Seminar" was one way of getting around prohibitions on weeknight film screenings on campus. At one such screening, he told a student reporter that he didn't care for Woody Allen but appreciated Peter Sellers. On that particular night, he was finally prepared for a viewing of *The Great Dictator*, with the radicalism of the 1960s in the rearview mirror. He played ragtime music for the students, who sipped on cider while watching the film. "We don't laugh enough," he told reporter Teresa Nicholas. He also decried recent cinematic ventures. "We are so conditioned to having slick movies—which may have no content."[18]

All These Worlds

The Barnard's Star planets were believed, even in the 1960s and 1970s, to be cold, large, and lifeless. Until the late 1970s, there were few inklings that a gas giant moon like Europa or Titan could possibly hold life, until the Voyager 1 and Voyager 2 probes opened up that intriguing possibility. And it seemed increasingly certain that there were planets beyond our solar system, meaning that our planetary system wasn't just a bizarre quirk of our Sun's formation that had never happened any place else, making the Earth exceptional. On the contrary, the universe was actually turning out to be a resplendent planetary wilderness, calling for humans to explore it through any means necessary.

That enthusiasm translated to the world of science fiction, of course, with Douglas Adams's novel *The Hitchhiker's Guide to the Galaxy*—in which Barnard's Star was a transit stop for the boorish Vogon fleets and their dreadful poetry—and Arthur C. Clarke's Rama series of books, all featuring the planetary system. Systems beyond our own are featured in sci-fi schlock as well, like an abortive *Battlestar Galactica* sequel called *Galactica 1980,* and *The Alien Encounters,* a strange 1970s made-for-TV mockumentary about the cover-up of probes from Barnard's Star visiting Earth that comes across as a lo-fi *X-Files.*

But perhaps the most famous use of a star and its planets comes from Robert Forward's Rocheworld series. In these books, humanity sets out to Barnard's Star using light sail propulsion and discovers a bizarre figure-eight-shaped contact binary planet.

A group of British researchers who wanted to make science fiction into science fact proposed a mission known as Project Daedalus, an attempt at reaching a good percentage of the speed of light. All stars are light years away from us—and most of the time, from one another. This means that any light we see from a star is its condition in the past. Every time we see Barnard's Star, which is six light years away, we see it as it existed six years ago (give or take the light months, days, and minutes that make it not an even 365.25-day per light year voyage).

But getting to those stars is a vastly different proposition. The Voyager 1 craft is leaving our solar system at 38,610 miles per hour, or nearly 11 miles per second. That's faster than the blink of an eye in human terms, but incredibly slow in cosmic terms. In all its years in space, the probe has managed to get just 18–20 light hours from home, meaning that it will be a long, long time before the future of *Star Trek: The Motion Picture* and its enigmatic sentient Voyager probe can come true.

Light travels at 186,000 miles per second. Voyager is going only 0.00005 percent the speed of light, in what suddenly seems like an incomprehensibly slow pace. In another 300 years, it will enter the last vestiges of our solar system in the Oort Cloud, and finally leave the solar system for good in 30,000 years. Voyager 1 will fly within a light year and a half of AC +79 3888, which is moving toward us, in 40,000 years.

All of this means that any conventional method of getting to Barnard's Star is out the window. Chemical rockets can't even get us to 1 percent the speed of light, and at 1 percent the speed of light, a voyage to Barnard's Star would still take 600 years, give or take. But if a method could be found to bump up to 10 or 20 percent the speed of light, then we could compress it into the framework of a human lifetime and see the planets for ourselves.

Well, from afar, of course. No one was seriously considering setting out on a human one-way trip, sight unseen. The British Interplanetary Society instead wanted to build a blueprint for an interstellar probe. And they wanted to use lots and lots and lots of nuclear explosions to do it.

The engineers and other scientists who conceived of Daedalus—which they quickly realized was probably out of the realm of near-future possibility because of the scarcity of helium-3—chose Barnard's Star explicitly because of van de Kamp's work. (The final report mentions the possibility of up to six gas giants there, showing how ideas of the system had grown in roughly ten years.) To get the right thrust, the probe would have to employ nuclear fusion, a wildly inefficient technology even today, and the primary source of helium-3 would have to be Jupiter, requiring a pioneering approach to human exploration on a scale we've never attempted. Thus, while the engineers at first attempted to show that building and launching an interstellar probe was feasible with the technology of the mid-70s, the need for fusion meant that the project would take closer to a century to pull off—all to reach the star in just 40 years. The probe would also need semisophisticated AI to accomplish its goals, something else that we are only just beginning to contend with. The craft would also require technology designed for on-board autonomous repair, something else that's a little out of reach right now.[19]

In the intervening time, several other mission concepts have come and gone, but each of them has used the basic template set forth by Daedalus, save Project Starshot, which would send multiple fairly unsophisticated small probes toward Alpha Centauri, powered by laser light sails. The sheer volume of probes would undercut the need for on-board repair.

Barnard's Star planets weren't, in other words, far from the public imagination, even if they weren't hospitable to life.

But what if they weren't there at all?

1
2
3
4
5
6
7
8
9

In the early 1970s, something was brewing upstate at the Allegheny Observatory outside Pittsburgh.

The observatory had barely survived the 1960s. Van de Kamp had worked to save the telescope, telling the University of Pittsburgh's chancellor, Edward Litchfield, that the telescope there was "one of the finest and largest of its type in the world" and that "the closing of the Allegheny Observatory would be a grave mistake." He expounded on its great contributions to astronomy, especially the work of astrometry.[1] The trouble began in early 1965, when the trustees of the university ordered the observatory closed, with Nicholas Wagman, then director, instructed to shut it down as soon as possible so that it could be demolished. As a result of outcry—from van de Kamp, several of his colleagues, the wider community of Pittsburgh, and more—the demolition was canceled, but the observatory was ordered to provide its own operating budget of $50,000 independent of the university.[2]

Litchfield was the force behind the funding woes—he had mismanaged an attempted expansion of the university, plunging the campus into debt. The chancellor was attempting to make the school a top-tier research university that could compete in the space age, and he believed that the observatory was "outmoded," saying, "The present equipment has long been obsolete. The University no longer can justify the expense of keeping it in operation particularly since the contributions we hope to make in this field of study cannot be adequately brought to fruition with the present installation."[3]

Litchfield was fired later in 1965, and the demolition was canceled when federal, state, and private funding came through.[4]

The observatory survived into 1973 (and is in fact still open today) in time for George Gatewood, a faculty member at the time, to make a peculiar set of observations. Pittsburgh and Philadelphia are on opposite ends of the same state and thus see much the same arrangement of sky. An observer there would see Barnard's Star in much the same way in either location. Gatewood was peering deep into the system and recognizing a problem: from his point of view, Barnard's Star wasn't wobbling at all; it was moving in a relatively straight line from the point of view of Allegheny. This flew in the face of a recent paper by Oliver Jensen and Tadeusz Ulrych, who had found that two planets weren't enough to explain their observations; instead, Barnard's Star likely had three to five planets.[5] (They later corrected this to three planets.)[6]

Like Sproul, Allegheny had several decades of plates on Barnard's Star, and Gatewood was working with a set of at least 161 plates created between 1916 and 1971. He and his collaborator Heinrich Eichhorn then compared these to 80 plates taken from the Van Vleck Observatory at Wesleyan University in Connecticut. Finally, they compared the first two sets of plates against Sproul's plates, but they kept running into the same result: Barnard's Star wasn't wobbling.

They ran and reran their results at the Strand Automatic Measuring Machine at the US Naval Observatory—a machine named for van de Kamp's former protégé. Once again, no significant deviation from a "straight" line was found. And when they attempted to use van de Kamp's orbital parameters for the companion object or objects, they found it didn't match *any* motion of the star.

"There are perhaps similar instances in the past, when astrometric investigations have suggested the reality of actually unreal things," Gatewood and Heinrich wrote.[7]

If the Barnard's Star planets weren't there, what did that mean for 61 Cygni or other members of van de Kamp's planetary pantheon?

A Flaw in the System

The first warnings had come from within Sproul. John Hershey was poring over old plates while measuring stellar motion and observed an unusual quirk. The star AC +65°6955 (Gliese 793) was on the move over the course of several years.

But the proper motion of Gliese 793 was well known. It was, in fact, frequently used as a reference star in the Sproul program, yet here was an unseen perturbation. This should not have been the case, but the proof was in the plates sitting before Hershey. One shift had been observed by the Sproul refractor in 1949, and another in 1957–58.

Those dates, eight-plus years apart, had a fairly humdrum significance at Sproul: they coincided with adjustments on the refractor during routine maintenance. Once these minute errors were accounted for, Gliese 793 showed no apparent wobble, and the positions of other stars appeared more "correctly."

The *objective lens* of the telescope, which is the one at the top facing the stars, appeared to have a particular defect when calibrated a certain way. Those minute repairs were enough to bump Gliese 793 off course. The most pronounced effect appeared to be on small background stars.[8] The 1949 adjustment to the objective lens had been the most notable. A new housing had been installed, and a new emulsion technique for installing the plates was put in place. Not coincidentally, one of the major shifts in the path of Barnard's Star had been noted in 1949. Hershey began to delve into other stars used by Sproul as reference stars. He compared them to plates from the Naval Observatory, Yerkes Observatory, in Wisconsin, and more, still finding subtle, unwarranted shifts in the positions of stars in the Sproul plates.

The astrometric binary program was in no danger. The shift wasn't significant enough to create a noticeable error in the trigonometric position between two stars and Earth. A system like Ross 614 had a significant enough interaction between its stars that the wobbling effect was real and pronounced. But observations like those of Barnard's Star, which stretched the limitations of the instrument to near the breaking point to capture less than the width of a human hair on the plates, could be greatly affected by any shift resulting from calibrating the objective lens.

Wulff Heintz, too, was noticing discrepancies in Sproul data. They appeared minute at first—binary stars that, when compared to plates at the McCormick Observatory in Virginia, didn't seem to have the "right" periods. Ever studious and serious, Heintz began examining some of the other discrepancies, presenting the possibility that many Sproul observations were in error at the International Astronomical Union meeting in 1974. The newest doubts about the Sproul data were coming from within Swarthmore.

Inner Friction

In 1975, there was a lot of pressure within the Swarthmore astronomy community. NSF cuts were looming. A visiting committee performing a sort of audit within the Astronomy Department and Sproul warned, "The discord in the department is a threat to continued funding."[9]

The line between Sproul Observatory—run by Lippincott—and the Astronomy Department—run by Heintz—was more pronounced than ever. Van de Kamp and Heintz, who had once had a sort of Odd Couple friendship—one the serious, dour German, and the other the jovial, showmanish Dutchman, devoted as much to vintage comedy and classical music as he was to the work of double stars—now held an open animosity toward each other. "Both van de Kamp and Heintz have not always been as discreet in remarks about the other as prudence would dictate it, and honest differences of opinion concerning a scholarly matter have been transformed into intense personal dislike," the committee found.

The friction was so well known that it had trickled beyond the campus community, all the way through the astronomical academy, and on up to the funders at the National Science Foundation. The final Project Daedalus report even made mention of the volley between van de Kamp and the wider astronomical community of the planets. The NSF was wary of funding such a caustic and combustible program with the principal leaders frequently and consistently at odds. Kaj Strand was among the members of the visiting committee.

A severe funding cut could cripple astronomy work at Swarthmore, which was then divided into two tracts of research: van de Kamp (not sitting well with retirement) and Lippincott continuing their work into astrometric binaries, and Heintz working on a completely different program regarding double stars. Heintz had also appropriated some of the astrometric funding into his own research, which the committee found could fall into questionable terrain, though Heintz had remarked that this appropriation had come from previous work in the Lippincott / van de Kamp program that had significantly cut into his own research time.

Within minutes of the committee's initial interview with faculty and staff, it was obvious that things were at or near a boiling point. Heintz steamrolled many of Lippincott's suggestions and concerns about the unhealthy state of relations between the observatory and the Astronomy Department. Heintz had his own views of how the observatory should be run, and none of them involved the ongoing astrometric studies of van de Kamp and Lippincott.

The core of the sour attitudes in the department seemed to be on Heintz, whom committee members said was "extremely proud of his scientific ability and tends to be contemptuous of work that he considers below his standards, and disdainful or impatient with those involved." They found that Heintz seemed to magnify perceived slights and had a "highly suspicious nature" along with a compulsively rigorous work drive. They also noted that Lippincott and van de Kamp "have contributed to some extent to the present dissension by some tactlessness and by a lack of understanding of, and concern for, Dr. Heintz's legitimate research and teaching objectives."[10] The committee found that Lippincott and van de Kamp were more amenable to changes in the department than Heintz had believed—albeit not without some rather personal input.

With all this considered, committee members were concerned that the NSF would not justify two expensive programs coming out of one department, so they made a few contingency recommendations. The first was that van de Kamp be fully retired, and the work shifted between Lippincott and Hershey, who both received general school money as part of their salaries rather than relying on grant-contingent funding.

Lippincott, at the time, was still unsure of her place within the college. Despite her years and years at Swarthmore, and her directorship of the Sproul Observatory, she still lacked tenure and had to reapply for the position every three years, giving her a feeling of instability, though she was one of the preeminent faces of women in astronomy at the time and a widely respected researcher. The committee noted that this likely contributed to the friction. Neither Hershey or

Lippincott had tenure, but Heintz did. As Heintz's complaints about the use of telescope time heightened, Lippincott worried that the decades-long work she, van de Kamp, Hershey, and others had undertaken would be thrown out. The committee didn't settle the question of who should necessarily be in charge of telescope time. Even while recommending tenure, the committee questioned whether she should, "in view of Heintz' undisputed superiority as an astronomer, be entitled indefinitely to have the last say in certain astronomical matters at the Sproul Observatory."

Hershey was in an unenviable position. He was an Astronomy Department faculty member stuck in the messy web between van de Kamp, Lippincott, and Heintz, all of whom he remained friendly with even as friction was at all-time high. He also had a knack for telescope repairs, which was helpful in view of the objective lens's difficulties. But Hershey and Heintz were under a lot of pressure in the department due to faculty shortages as well, which were leading some budding astronomers to seek out coursework at other nearby schools.

In the end, the committee report recommended that Heintz be granted full right to the use of observatory facilities and that Lippincott and van de Kamp not interfere. Further, Heintz and Lippincott should work out funding matters between them. And, though Lippincott would have all final say in observatory matters, the memo suggested that she stop dismissing some of Heintz's suggestions out of hand, instead writing explicit reasons for denying his requests. The same went for Lippincott's suggestions to the Astronomy Department, where Heintz was to grant the same degree of respect. Finally, the committee suggested that unless van de Kamp's research had specific funding for a specific program, he should take a back seat in all matters of both the observatory and the department.

Ultimately, Lippincott was granted funding to continue the astrometric binary program for another two years, from November 1976 to October 1978. Hershey was also granted funding under the program, but Heintz was not. A 1979 technical report to NSF still attested to evidence for two planets at Barnard's Star, as well as the detection at

Ci 18, 2354/Gliese 687.[11] On the Barnard's Star situation, Strand told Steven Dick and David DeVorkin,

> I'm neutral on that point. Not for reasons of my relation with van de Kamp, but simply because he has a point, that he has done so many observations of it over such a long period of time, and we have attempted to do it at the US Naval Observatory with the 61-inch and have not been able as yet to say yes or no, because it's a tricky problem. You see it is not only a matter of measuring Barnard's Star against a number of background stars. For these background stars, you have to know their proper motions very accurately, and this makes it very difficult. You cannot measure the Barnard Star's perspective acceleration if you don't know this about the background stars.[12]

Yet, results that indicated two or more planets at Barnard's Star flew in the face of other available evidence, and by 1976 Heintz was making his dissent with the discovery known, publishing a paper in the *Monthly Notices of the Royal Astronomical Journal* stating that there was no evidence for a substellar companion at Barnard's Star. Heintz found that most pre-1945 evidence had to be thrown out because of its perceived weakness. Once that evidence was eliminated, there were few, if any, significant derivations of signal from noise.

To everyone but a small handful of true believers, it seemed there weren't one, two, three, or five planets at Barnard's Star. There were zero.

The Dominoes Fall

Shortly after Gatewood had refuted the discovery of Barnard's Star planets, he set his sights on another system: Lalande 21185, the site of the second discovery at Sproul after 61 Cygni. Gatewood took 143 exposures of the star from the Thaw refractor at Allegheny, a photographic workhorse that van de Kamp had worked hard to save a decade prior.

Applying the same methods he'd used to disprove the existence of the Barnard's Star planets, he also failed to find evidence for the planet or star-planet at Lalande 21185.[13] This ruled out at least two of the handful of planetary discoveries at Sproul—and significant doubt was being cast on the other discoveries there. In fact, by 1975, he had

5.1

Sproul Observatory staff (left to right): Wulff Heintz, unknown, Peter van de Kamp, John Hershey, and Sarah Lee Lippincott, looking more like a strange band than an astronomy faculty

Courtesy Friends Historical Library of Swarthmore College

significant doubts about 61 Cygni as well, and he suggested that the tentative detections at EV Lacertae (BD+43 4305) and Epsilon Eridani needed a wider net of data to stand up to scrutiny. "The probable cause of the long list of astrometric disappointments is unstable astrometric instrumentation," Gatewood wrote. "However, instrumentation and techniques do exist which, if directed to the task, could detect the effects of Jovian mass planetary companions on their primary stars."[14]

Strand told Dick and DeVorkin, "61 Cygni didn't have the companion as assumed at that time, which was unfortunate, but we didn't know enough about the optical performance of refractors or how accurate they were. We all thought that they were the ultimate in accuracy that we could achieve in astrometry." He still, in that interview, insisted that something might be there.

Gatewood wasn't necessarily setting out to disprove the results of van de Kamp and Lippincott (and Strand)—though, indeed, that was the result. He wasn't squeezed for telescope time by what he saw as a frivolous pursuit of phantom planets. He was instead on much the same quest as Lippincott and van de Kamp—to find invisible worlds. Allegheny would soon undertake a Sproul-like astrometric program, looking for minute deviations in the paths of nearby stars indicating unseen companions. Gatewood wanted to find planets—but they weren't readily revealing themselves. In 1978, he published a paper tentatively suggesting evidence for a brown dwarf in the Altair system, but it was not widely publicized or accepted, and Gatewood himself would later refute the results.[15]

Gatewood was as ferocious as van de Kamp in finding these worlds but had the benefit of hindsight and a relative lack of accumulated hubris from a four-decade hunt. When continued parallax studies failed to yield results, Gatewood began suggesting that astrometric searches for planets should take place in space, above the disturbances of Earth's atmosphere.[16] He didn't find planet hunting to be distracting from more serious astronomical pursuits, as Heintz had suggested. Gatewood was demanding that *more* serious gravity be applied toward this sort of planet hunting.

Homefront Battle

In a 1978 paper, Heintz continued to poke holes in van de Kamp's worlds, refuting the existence of the *star-planet* at Epsilon Eridani and pointing out another discrepancy in the Barnard's Star data. There was an identical derivation in Gliese 809 (BD +61°2068), a star similar to Barnard's that showed up in several of the same exposures. It is incredibly unlikely that two similar objects would have planets with identical periods that appear to orbit in the same intervals at the same time.

Heintz torpedoed Strand's work on 61 Cygni in the paper, calling the original material "quite weak" compared with subsequent observations, then dismissing the work on Lalande 21185 out of hand as well. Ci 2354/Gliese 687 was gone. Sproul Observatory had discovered a handful of planets, but in the blink of an eye, they were all wiped away by a fastidious researcher who wanted to cast aside the poor quality of the data in light of the recently detected errors.[17]

As noted in a previous section, a 1978 observatory report, the same year Heintz published his paper, pushed forward with the original Barnard's Star hypothesis—all in data gathered by van de Kamp.[18] It also offered up a new possible planet or brown dwarf around Groombridge 1618.[19] But a 1980 update seemed to agree with Heintz's opinion that there were no planets around Lalande 21185 or Epsilon Eridani. The researchers had found evidence by 1980 of a brown dwarf (although still not referred to by that name) orbiting CC 1228/Gliese 806. But Groombridge 1618 and CC 1228 weren't adding on to a pantheon of planets coming out of Sproul.[20] They were, instead, seeming to mark a clean slate for the observatory, as van de Kamp's influence began to decline more and more.

As disagreeable as Lippincott and van de Kamp had found Heintz, he was a thorough enough researcher that they eventually had to concede that he was right.

Or at least Lippincott did. Van de Kamp never yielded such ground.

It's easy to paint Heintz as a villain, but a 2001 interview with the *Swarthmore Bulletin* paints a much, much different picture. Heintz was a methodical perfectionist who wanted a similarly methodical precision to the data. He had his inclinations that something was off with the data as early as 1970. Van de Kamp had asked him to gather additional observations, but Heintz wasn't able to obtain previous data from van de Kamp, who kept the Barnard's Star plates closely guarded. Once Heintz was able to get a hold of them, he was troubled by their deviations from other observatories, as well as the purported movement that, when paired with uneven exposure of the plates, made any tugging effect seem miniscule.

Heintz recalled that this was around the time that he and van de Kamp had begun to have a sort of schism in their friendship, which turned into outright animosity. When Heintz tried to raise the issue with the college, it kicked off a war of words that rippled all the way to the visiting committee, before Heintz had even committed his rebuttals to paper.

"I was denounced among his friends—including top administrators—as a nasty character and probably mentally disturbed," Heintz told journalist Bill Kent. "I was told that I should do nothing about it, and that his observations would eventually be confirmed. This I could not believe because, with the variations in the exposures alone, there did not seem to be enough to make any conclusion either way."[21]

Heintz, ever thorough, felt he was simply fulfilling the work of a scientist—astronomy, in all its grandeur and heavenly vistas, is as much a matter of numbers as it is objects in the sky. An observatory Sproul's size was most effective in helping create a map of the nearby heavens, rather than entering the planet hunting business—at least in the view of Heintz, whose conclusions were that you couldn't gather planetary data from astrometric photography, no matter how much you tried to push a ground-based instrument to its limits. The width of movement was a fraction of a hair, just enough for a lens to be maladjusted, or any number of other variables to throw off a signal ever so slightly, especially when measuring by hand or analog instrument.

"I can tell you . . . I did not want to embarrass a colleague," Heintz told Kent. "There was no satisfaction in it. I took refuge in my own studies, in my teaching and with the telescope. It's still quite a useful instrument."

A One-Man Quest

Even though Gatewood's actions started all the trouble, van de Kamp seemed to harbor little ill will toward him and even said that Gatewood had published his research in a "very fair way." But van de Kamp thought Gatewood lacked the sheer volume of data that Sproul had amassed over the years—van de Kamp was working with 20 times more data, for example—though as Hershey and Heintz were pointing out, the Sproul data had significant flaws. Van de Kamp told David DeVorkin, "The only thing that was at fault there was that some people took this as indicating that the Sproul observations were no good because they were not confirmed, and there was some unpleasant personal stuff involved there which I'd rather not go into."[22]

Van de Kamp raised the possibility to DeVorkin that it was a mild conspiracy within astronomy, that Gatewood was directed to discredit him. Van de Kamp felt that the person behind this—whom he would not name—wanted the onus to fall on him to collect more data on Barnard's Star, which was what van de Kamp was doing at Sproul as much of the rest of the campus community moved on. "Well, I have been told that Gatewood was told that 'Well, as long as you cannot confirm van de Kamp, make the most of it, that you disconfirm him,'" he relayed to DeVorkin. Van de Kamp expressed skepticism about the perceived accuracy of digital computing versus traditional hand calculations and plate comparison methods. When DeVorkin asked him why he continued on with the work, van de Kamp responded, "Because basically I'm terribly stubborn." He also suggested that Kaj Strand was seeing perturbations at the USNO. DeVorkin and Dick's interview with Strand a few years later revealed Strand's sympathy even in his desire to stay neutral.[23]

In these sessions, van de Kamp was fairly even handed about Heintz the scant few times he mentioned his colleague, which was mostly regarding Heintz having been passed over for the Sproul directorship.

Early Warning Signs

George Gatewood wasn't the first to sound the alarm on the Sproul refractor's small—but when working with small perturbations, pronounced—issues. In addition to the noted calibration issues with the objective lens, astronomer Phil Ianna first noticed a few thermal problems with the lens in 1962, which he wrote about and published in *Vistas in Astronomy* and the *Astronomical Journal*.

Astronomy is no stranger to thermal issues—especially with larger lenses on refractors. Changes in temperature and humidity can cause expansions and contractions on the equipment when exposed to these elements. Ianna's initial paper—published before van de Kamp was widely publicizing his putative planetary pronouncement—found that images could be shifted by as much as 24 microns, depending on weather conditions.[24]

In a 1975 NASA Ames astrometric meeting, Ianna reiterated the thermal problems of the Sproul refractor—and proposed the need for more reference stars to correct for these distortions in plate materials. He pointed out that thermal deformation of the lens component in the long-focus refractor is an important contributor to the errors in astrometric positions. "The cooling of the objective is non-uniform and lags behind the nightly variation in the ambient temperature with resulting changes in spherical aberration, coma, astigmatism, and optical distortion."[25] A 1957 adjustment had improved on the 1948 adjustment and reduced some of these errors, but Ianna still voiced his concerns about the lens. The minutes of the 1975 meeting record the conclusion that various plates of the same region from different observatories would need to be compiled and compared, one against the other, to rule out these small distortions.

Ianna has a small—but unusual—position in this story. Though he may have found some reasons that the Sproul lens was unreliable on small scales, he believed in van de Kamp's planets. In fact, he continued to pursue them, even after they'd fallen out of favor. In 1980, Ianna and Laurence Fredrick, then both of the McCormick Observatory at the University of Virginia, gave a conference presentation suggesting that there *might* be something in the Barnard's Star system . . . but that the results were ultimately inconclusive.[26]

A thorough analysis in 1985, however, put to rest any hope for the planet's existence. McCormick didn't see the planets' movement.[27] Nor did Allegheny or the USNO. Only the Sproul plates showed movements, and some faculty members there didn't see them, either. They were, in all estimations, in the eye of van de Kamp and van de Kamp alone. At subsequent American Astronomical Society meetings in 1995 and in 2001, Ianna continued to see nothing out of the ordinary around Barnard's Star.

Later Years

A 1977 paper showed that van de Kamp was, indeed, persevering through the perceived adversity. He writes in his conclusion, "There is the danger of either under-discussing or over-discussing data which are not much above the threshold of attainable accuracy. There is the psychological factor of being confronted with a discovery, and one may hesitate to take the initiative to follow through the signal."[28] The paper is defiant and rather unusual—flowery prose, religious iconography, and scientific data unevenly meshing. He expects that a complete orbit will be witnessed via wobble by 1984, vindicating his results, but also includes religious art and sections of music in the paper. If his data were just at the edge of noise versus signal, van de Kamp expected the signal to prevail over time.

It never did.

He stayed at the Sproul Observatory through the end of 1976, when he retired from the institution for good. By 1981, he had moved back to the Netherlands. In 1986, he published his final academic paper.

In that decade span, van de Kamp remained focused on vindicating his work. In 1982, he presented at the 160th AAS meeting on his continued Barnard's Star work, this time as a researcher with the University of Amsterdam. His talk was scheduled between Gatewood upholding his own results and Lippincott giving a status update on the search for other planetary companions. His explanation for the orbit of the planets becoming more chaotic was that they were tilting far above the ecliptic of Barnard's Star and at strange oppositions to each other.[29] Other updates in his searches began to appear in the German journal *Astronomische Nachrichten*, rather than the *Astronomical Journal* that had traditionally carried his work. In nearly everything he published, he would either directly or obliquely mention his work on Barnard's Star, even in a brief biography of unseen companion pioneer Friedrich Wilhelm Bessel. Van de Kamp found a way to conclude that biography with a tenuous discussion of his Barnard's Star discovery.[30] A few years earlier, in 1981, he released the book *Stellar Paths: Photographic Astrometry with Long-Focus Instruments*, which was meant to be an update to *Principles of Astrometry*. Instead, it takes a winding path toward discussing Barnard's Star. Most contemporary reviews of the book criticized the Barnard's discussion as a potential weakness, especially as it stretches the limits of what astrometry could do at the time.[31]

His final publication, *Dark Companions of Stars*, took a similar path. It opens with a biblical quotation, John 20:29, which says "Blessed are they that have not seen, and yet have believed." The manuscript makes an interesting inference—that someday, planets would be discovered by the way they dim their star's light.[32] This has, in fact, come true. Referred to as the *transit method*, it is one of the principal techniques currently employed for finding planets.

Most other Sproul planet candidates, save the massive candidate at EV Lacertae, seem to have been swept aside, but van de Kamp concludes the book, again, with a defense of his Barnard's Star observations—as if they should be regarded alongside stronger findings like Ross 614B. That work at Sproul—identifying very low-mass stars orbiting a larger star invisibly—has been vindicated over time and remains an important contribution to the field of astronomy.

Yet try as he might, van de Kamp could never manage to bring the Barnard's Star planets back into the pantheon of known planets. At the time *Dark Companions* was published, researchers were only beginning to reckon with the possibilities of truly finding planets in our galaxy. Van de Kamp's run of not-quite discoveries was a dress rehearsal for the same turbulence that would plague the era until the first real exoplanet emerged from decades of retractions, false positives, tentative quirks, not-followed-up-on observations, and other anomalies left in the dustbin of astronomy.

1
2
3
4
5
6
7
8
9

The exit of van de Kamp marked a quieter time at Sproul.

Sarah Lee Lippincott still sat at the helm of the observatory. Although possible that she and Heintz never saw eye to eye, the bitter, prolonged infighting had largely quieted. Rather than dwell in the past of the program and continue to push for a planet that just wasn't there, Lippincott and Hershey moved on with the hunt for unseen companions around nearby stars. This time, they had the benefit of hindsight. Though they hadn't completely, internally, ruled out planetary candidates around BD +43 4305 and Epsilon Eridani, Lalande 21185 was eliminated from the list of candidates.

They were also contending with a whole host of new possible worlds, including something peculiar at Ci 18, 2354, more commonly known as Gliese 687. They weren't entirely sure what was there, and they placed it anywhere from a six Jupiter-mass planet to a tiny, tiny red dwarf. The team was also working hard to find a companion around Chi1 Orionis, a Sunlike star with a large wobble, which ended up being the result of a small perturbing star—the second such invisible star found at Sproul in the mid-1970s. (The first was PGC 372.)[1] There was additional evidence for a star in the Wolf 1062 system, which was shown to be absolutely true (and just a hair above Sproul's mass estimate.)[2]

Gliese 687 never yielded the predicted planet, though in 2006, a Neptune-sized world was discovered there.[3] Detecting something as small as a Neptune-sized world was unlikely in 1977 by any available means. The paper from Lippincott announcing its discovery was somewhat notable in that she'd consulted Heintz on possible observational changes, disregarding some of the enmity between the two.[4] (It's possible that the threat of losing NSF funding was enough to scare the institution into putting on a good public face.)

Van de Kamp didn't quite spend all his time embroiled in the Barnard's Star affair. A 1977 letter from Lippincott to Laurence Fredrick—written a bit after his departure to the Netherlands—asks Fredrick, then at the McCormick Observatory, for more information on Groombridge 1830.[5] There appeared to be some strange activity at the star, but there wasn't a planet to be found there, either. Instead there was evidence of a new type of stellar mass ejection (often called a flare).

At this point in the Sproul observing program, late 1970s, Lippincott was largely carrying on van de Kamp's initial 1937 program. It made sense. She'd been with the program from near its inception, and despite a few planetary hiccups here and there, it had been successful in finding unseen binary stars and gleaning other information about our nearby stellar neighborhood.

Hershey was there too, at least part time, and Heintz continued his hunt for double stars. Michael Worth was a full-time assistant, while Mary Jackson and Ruth Kennedy provided part-time support. And Elliot Borgman was there in an apprentice-type position, working under Lippincott. It was a skeleton crew working on a skeletal program at a small college.

Van de Kamp's career had wound down with a sour note, an enormously respected researcher making an error and digging in his heels on it. But Lippincott wanted to persevere with the benefit of hindsight and look for invisible stars, planets, and the objects in between with more and more certainty. She proposed an object of 20 Jupiter masses around the star CC 1228, a.k.a. Gliese 806, and, at the same time, correctly identified a small binary star to Wolf 922.[6]

But Gliese 806 has a key difference from many of the worlds found at Sproul: its companion has never been disproved.

In 1989, astronomers Geoff Marcy and Karsten Benitz compiled a list of stars that showed unusual radial velocity measurements or had otherwise anomalous motions. Marcy, largely known for his work at UC Berkeley, was an important figure in early exoplanet hunts. Although he didn't find the *first* exoplanet, he found the second. And third. In fact, he found a great number of early exoplanets—70 of the first 100 discovered. He also had a hand in several projects, like Kepler, that have found planets by the *thousands*. But Marcy's tremendous impact within astronomy is impossible to separate from the sexual harassment scandals that have plagued his career in the last few years, allegations that stretch back to the 1990s.

Marcy and Benitz took a sample of 70 nearby stars—which would well overlap with Sproul's studied stars—and found at least four planet or brown dwarf candidates at Glieses 380, 521, 623, and 806. Three of these four were also suspected by Lippincott and Hershey to have planet-ish objects.

Gliese 380 is better known as Groombridge 1618. In 1978, Hershey and Borgman made the case that an object between 3 to 12 Jupiter masses was in orbit around that star, which is a bit smaller and dimmer than the Sun (but larger than a red dwarf.)[7] Marcy and Benitz proposed a 4 Jupiter-mass object in a 122.5-day orbit. But to date, those observations haven't been followed up on. Groombridge 1618b could be sitting out there in its short orbit, just waiting to be discovered.

Gliese 623 appears in Sproul literature as CC 20,986. Lippincott's observatory notes give a wide range of matches but zero in on 0.06 to 0.08 solar mass, or between 62 and 83 Jupiter masses, all at the boundary of the brown dwarf/star divide. Marcy and Benitz place it at 0.08 solar mass, a figure later restated by Marcy and coauthor David Moore after more intensive studies.[8] Marcy and Moore also credit Lippincott and Borgman's work on the object. The object was finally confirmed in 1994 by the Hubble Space Telescope, which looked at the Gliese 623 system with enough precision to make out the space between the two objects—and to observe one of the smallest stars ever discovered to that time. Astronomer Natalia Shakht estimated the mass—just via astrometry—to be 9 percent of the mass of the Sun in 1995.[9] Precise mass measurements in 2007 confirmed it as a star at 11 percent the mass of the Sun.[10]

Gliese 806 was projected by Lippincott to sit at about 20 Jupiter masses. Marcy and Benitz's work placed that value at closer to 11, which would take it from a brown dwarf to a planet by some definitions. Gliese 806b, too, may be awaiting discovery.

There's also something of a mystery hiding in the Sproul correspondence. A 1980 letter from Lippincott to Laurence W. Fredrick at McCormick Observatory requests more information on BD+36 1638, a star about 38 light years away. Lippincott believed that something was in orbit around this star in a period of 40 to 80 years, but she wanted McCormick plates to be certain.[11] The star does have a known visual binary—Ross 989—but isn't known to be attached to any star. Like EV Lacertae, it was known to be a flare star.

Whatever was there, it never seemed to go to press. Perhaps it had a signal-to-noise ratio that was too close for comfort. Maybe the plates she requested from McCormick showed that the same instrument error that had stymied research time and again at Sproul was up to its old tricks again. But a weird twist in this little footnote comes in the form of a 2008 paper that used adaptive optics to search for faint companions to stars. Adaptive optics are telescope instruments that reduce atmospheric distortion, partly block out starlight, or use some other method to directly image a previously unseen companion to a star. This 2008 paper showed a visible companion to BD+36 1638, which it refers to as GJ (or Gliese) 277A. (Standardizing star names is, to say the least, a chore.) The researchers detected a very slight radial velocity change in the star and another small object hiding in the field of view. It may just be another star hiding out in this planetary system in a weak binary orbit.[12] The parent star—a fairly humdrum red dwarf—certainly hasn't made itself stand out for follow-up investigations. But it's possible that Lippincott had caught a hint of this little tug somewhere.

Free-for-All

After a few Barnard's Star–sized missteps, the real groundwork was being laid for planetary discovery in the 1970s. But Sproul was seeing the limitations of ground-based astrometry and wanted to continue the astrometric work about 350 miles away.

Straight up.

The Hubble Space Telescope had been on the drawing board decades before it launched aboard the space shuttle *Discovery* in 1990. The first concrete plan for Hubble arrived in 1968, with roughly the specs of the instrument that's still up there today, nearly 30 years after its launch. Actual preparations began in the 1970s, with a planned launch in the early 1980s, though like all good NASA projects, it ended up long, long delayed.

But this meant that, by the mid-1970s, the astronomy community was already buzzing with the possibilities of what a space-based observatory could do. There had been a few space telescopes by that time, including the notable Orbiting Astronomical Observatory fleet (or the two of them that succeeded anyway), but none as powerful as Hubble.

It was still referred to as just the "space telescope" among astronomers and in other space circles at that time. In 1977, Lippincott and Hershey prepared a proposal to coordinate the space telescope with Sproul's observing program and suss out the finer details on stars that were beyond the reach of the equipment at Swarthmore.[13] And in 1980, an astrometry working group, which included Gatewood, put forth the idea that space-based astrometry was the best way to detect an Earth-size planet, which would, according to the group, cause a 3 micro-arc-second disturbance in its star.

A copy of this group's paper in the Swarthmore Archives has several notes from Lippincott written on it in red pencil. In the sentence "Planetary detection is one of the most exciting new fields of modern astronomy," for example, she has crossed out "new" and written "since 1937" in the margins.[14] There's a contrast of opinions in that simple pencil stroke. On one side, you have astronomers who felt that the field was just beginning. In the minds of the wider astronomy community, there were no data points on extrasolar planets—no Saturn-like world hanging on the outskirts of Barnard's Star, no star-planet at Epsilon Eridani, or a third component to 61 Cygni weaving its way awkwardly through the system. Astronomers as a whole were not working against a backdrop of 43 years of planet-hunting institutional memory. At this point in astronomical history, most outside Swarthmore felt that the hunt for planets had moved on and that astrometry needed to come into its own to really, truly find planets.

Ironically, this working group report concluded that astrometry was best suited to find extrasolar planets. Yet most planets have instead been found by the transit method or radial velocity; none has ever been conclusively found through astrometry. But if the worlds pointed out by Marcy and Benitz ever bore fruit, then Sproul would have one of the first success stories, moving the planetary discovery calendar back a decade-plus.

The idea of using the space telescope to directly follow on to Sproul's work never quite came to fruition. In the late 1970s, Hershey began working on photometry with the University of Maryland. Though he came back to Sproul for one year, he eventually left for the USNO, where he worked on ground-based follow-up to Voyager 2's encounter with Uranus in 1986. While there, he was also part of a team developing an interferometer to help further refine astrometric measurements. He subsequently landed at the Computer Sciences Corporation within the Space Telescope Science Institute (STScI), the parent organization of Hubble, where he spent the rest of his career. Thus, though Sproul's program was never fully realized on the orbiting telescope, Sproul still played a hand in its work.

In 1997 Hershey coauthored a paper with Al Schultz and several other STScI astronomers noting that Proxima Centauri, the nearest star to Earth, at 4.2 light years, had some kind of companion at 0.5 AU, something that definitely wasn't a background star.[15] The discovery came and went with little fanfare at a fairly confusing time in exoplanet research. George Gatewood, at the time of the discovery, told *New Scientist* reporter Gabrielle Walker that he wasn't convinced the discovery was correct, but he was "intrigued" by it.[16]

A lot of strange *ifs* about the potential Proxima Centauri companion added up to a unique puzzle. In 1997 it was estimated to be a likely brown dwarf, but there was no wobble, which should have been easy to detect for an object that big that close with technologies available at that time. It also appeared to have an eccentric orbit, which Gatewood told *New Scientist* was an unlikely scenario.

But in 2016, astronomers announced the discovery of a planet in a close-in orbit around Proxima Centauri, one that was a bit more massive than Earth but likely rocky and in the habitable zone of the star. There was also tentative evidence for a second planet.[17] That same team later drew out further details about the Proxima Centauri system. For instance, they detected two different belts of dust, the sort of stuff that provides the building blocks of planets, moons, and other objects in planetary systems. It might be something akin to our asteroid belt.[18]

The inner belt—which is located at about 0.4 AU—could be shepherded into place by a planet at 0.5 AU. (They also reported another planetary candidate at 1.6 AU, meaning there could be three or more planets in the system.) A planet—potentially of gas giant mass—at 0.5 AU could be a big boost to the 1997 detection, which seems to have fallen through the sands of time. In an interview I conducted for *Discover* magazine, Itziar de Gregorio-Monsolvo, a coauthor on the 2017 dust belt discovery, told me, "Our ALMA data reveal also an intriguing faint compact source at a distance less than 2 Astronomical Units from the star and that could be interpreted as a ring of dust surrounding a giant planet 100 times more massive than the Earth."[19]

Near Misses

While some candidates at Sproul have never been disproved—those few potentially corroborated by Marcy and Benitz—others quickly fell away. For instance, in 1979, Lippincott made the case for a 2 or 4 Jupiter-mass planet around the star EV Lacertae (BD +43 4305). But so far, no planets have been found there. What does appear is a series of active flares, some of the largest known anywhere, which would almost definitely throw off the flux of the star.[20] Lippincott, however, went as far as speculating that the planet could be a catalyst for these megaflares.[21]

EV Lacertae is a fairly young star, and its flares are so powerful that a 2008 event blasted the space around it with 1,000 times more energy than we've ever witnessed in such an event. Over the past century of observing red dwarf stars, astronomers have noted that young ones tend to be some of the most temperamental stars out there, flaring out ejections in all directions. That means that while stars like Barnard's and Proxima Centauri are small, they pack a powerful punch. The resulting radiation can rain down on any planets around them and potentially rip away their atmospheres, one of several reasons they might not be so friendly to life.

There's also the case of Stein 2051, which van de Kamp was working on toward the end of his time at Sproul. This object brought Strand and van de Kamp back together on the hunt for a planet or brown dwarf. Data from Sproul and the USNO suggested a most unusual system: a red dwarf in a binary pair with a *white dwarf*, which is the husk of a star like the Sun after it reaches the end of its life as a "normal" main sequence star. But Strand seemed to think, in a paper published in his last year at the observatory, that the system had a 20 Jupiter-mass object as well, ever so subtly throwing off measurements of the pair.[22]

Strand relied heavily on data from his old stomping grounds, including those gathered by van de Kamp, Lippincott, and Hershey. Van de Kamp, in correspondence with his old friend Strand, was eager to find out more about the putative planet. Strand's search had been taking place for a few years, with van de Kamp catching wind of an anomalous signal around the red dwarf as early as 1974, when he was putting together *Unseen Companions of Stars*. "Both you and I should urge Flip [to] see to it that this object is attended to this current season," van de Kamp wrote Strand. "I should like, of course, to include this object in my study and listing."[23]

The letter was handwritten, with none of the formality of a typewritten letter dictated to an assistant. It came from the University of Amsterdam, van de Kamp's new perch. Van de Kamp was, of course, in the middle of the Barnard's Star fallout and perhaps eager to see his former student succeed as his own planetary pantheon fell out from under him. Strand had spearheaded the first success, all the way back in 1943, while fighting for his adopted country. But that success had faded over time.

At that point, in 1974, before anything had been published about the stars and their companion, Strand wasn't quite so ready. "It would be premature to give data on that system until I have a better idea of the period of perturbation," he wrote. "Hopefully this year will show a turn around. However, there is not guarantee for this. Moreover I have published nothing about the system as yet, which I think should be done first." Strand also expressed admiration for Hershey's work and demeanor.[24]

A 1980 *Philadelphia Inquirer* profile of Lippincott seems to imply that the Stein 2051 companion is a given, alongside CC 1228, with a more ambiguous mention of Barnard's Star.[25] Lalande 21185, Epsilon Eridani, 61 Cygni, and other discoveries bear no mention. That same year, Hershey wrote to David C. Black at NASA Ames, listing a few more planets as givens. Barnard's Star was still included, despite Hershey's role in its downfall, perhaps out of respect to van de Kamp's ongoing work. Stein 2051 was of course there, as were Ci 18, 2354 and BD +43 4305. But Hershey mentioned that the Barnard's Star signal was weakest. "In that sense, the others are stronger cases for the presence of planetary or substellar masses." Epsilon Eridani and Lalande 21185 were excluded entirely, with Hershey explaining that they "do not show a cyclic pattern."[26] Hershey also wrote that BD +43 4305 was giving off different signals than previously noted. He indicated that Sproul had been error correcting its previous plates, but that doing so was roughly exhausting the capabilities of the plates as they were. "We believe that new instrumentation is needed in the planetary search field[,] and development should be supported," he wrote.

But by 1989, Strand was questioning his own conclusions. After further work, he had determined that no third object was to be found at all. Further exposures between his original publication and 1987 had made the case for a Stein 2051 planet so weak as to disappear, as had happened before with so many planets adjacent to Sproul.[27]

Today, Stein 2051 is considered important because of an event showing the weird physics at play within a white dwarf when the white dwarf passes in front of a background star.[28] Giant stars create black holes and neutron stars, but Sun-type stars leave behind *degenerate* forms of matter, remnants with electrons in bizarre configurations. Ionic matter swims in an electron field as the destructive engine slowly burns out over the course of few trillion years.

As Strand closed the book on Sproul's work—with the Sproul program left with little to show aside from a few objects neither confirmed nor retracted—the new generation was ready to build on its lessons. Interestingly, Ci 18, 2354/Gliese 687 does have a planet, but it's definitely not the 10 Jupiter-mass object reported by Lippincott. In 2014, astronomers at the Lick Observatory reported finding an 18 Earth-mass planet around the star, which is the mass of Neptune.[29] Gatewood had followed up on Lippincott's object in 1988 but found no astrometric signal. Nonetheless, he still recommended the star as an ideal place for a planetary hunt.[30] But the Lick paper indicated that the region of the sky where Gliese 687b was found had gone unexplored in recent surveys. The planet was discovered by radial velocity, which can find planets today around the size of Earth, whereas astrometry has largely been used in the realm of gas giants.

Another Chapter Closed

Two small objects in the solar system are named for the Sproul alumni. The 1965 van de Kamp is a seven-mile asteroid residing in the asteroid belt. It was named for Peter van de Kamp in 1979, reflecting the esteem he still held in the community. The 3236 Strand, discovered in 1982, is a twelve-and-a-half-mile rock that wavers into the inner reaches of the asteroid belt at times.

Despite Lippincott's contributions to the field, she has no asteroid to call her own.

The 1970s had brought great changes to Sproul—and the same was definitely true for Lippincott's personal life.

Dave Garroway was an early anchor of *The Today Show*, NBC's venerable morning program. The gregarious, seemingly happy-go-lucky anchor held court for viewers each morning throughout the 1950s. He had been with the network since 1938, when he had entered the page program at $16 a week. He also hosted a radio program and another NBC telecast, *Wide Wide World*.

But when his wife committed suicide in 1961, Garroway left NBC behind. There are conflicting reports on why—some sources attribute it to his wife's death, but others suggest that network reshuffling led to his departure. He tried his hand at a few other programs, including

Exploring the Universe on NET (which is now PBS). His career never fully recovered, but he managed to keep a low-key presence throughout the 1960s and 1970s.

Astronomy was one of Garroway's greatest interests. He'd studied it in college at Washington University before embarking on a career in television. He even bought a 10-inch refractor from Beloit College, building an observatory at his home in Long Island.[31] Garroway eventually sold the telescope to Keystone College, a liberal arts school just outside Scranton, Pennsylvania.[32] As he made his way through television gigs and relaunches, he also began an extensive tour of observatories across the nation. In 1975, that led him to meet Sarah Lee Lippincott.

Flip had, in all her time at Swarthmore, stayed devoted to her profession and hadn't married. This was the kind of independent streak that Sandra Faber had admired. But when Lippincott and Garroway happened to be on the same tour of Russian telescopes, they struck up a romance. "That was a very interesting match. I once visited them in their home together," Faber told Lightman. "It was lots of fun interacting with the two of them. But she was a very lively, vital person, clearly a very intelligent person, the first woman I had ever seen making it on her own."[33]

The couple's 1980 wedding ended up in, of all places, the *National Enquirer*. The bride was smiling from ear to ear. The couple had intended to wed in 1977, but a heart surgery had postponed the date. "I've waited a long time to find the right man," she said.[34] It was Garroway's third marriage and Lippincott's first.

Lippincott began winding down her role at the college, working part time while commuting between her home and Sproul by bike. Her final days in an official capacity at Swarthmore and Sproul came in the summer of 1981. After this, she became a research astronomer with the college.

At this time, Garroway had another heart surgery and found his health in decline, which also exacerbated his long-standing struggles with depression. He made an appearance on *Today* to celebrate the

30th anniversary in January 1982, promising to come back for the 35th anniversary. But at this time, he was slowly deteriorating physically and mentally.

On July 21, 1982, Lippincott had breakfast with Garroway before running out on an errand. She left the house at 8:45. A housekeeper came by at 9:30. In the interim, Garroway had grabbed a shotgun and killed himself with a single shot to the head. Writing in the *New York Times*, reporter Sally Bedell said, "Mr. Garroway's son, Michael[,] said that his father 'had been suffering from post-operative complications following open heart surgery and we were extremely surprised at the turn of events. We believe he unfortunately succumbed to the traumatic effects of his illness.'"[35] A private funeral for family and friends was held, with Michael throwing a public concert to commemorate his father.[36]

"You never know what demons work," *Today* producer Steve Friedman told United Press International. "The thing that makes these people so special may be the very thing that destroys them."[37] Garroway's children and Lippincott helped set up the Dave Garroway Laboratory for the Study of Depression at the University of Pennsylvania in his honor.

By the next year, Lippincott had published her final academic papers. One of them purported a very small star or very large brown dwarf around 36 Ursa Major—a claim never really refuted nor entirely proved.[38] Hershey served as an interim director at Sproul before John Gaustad took over in 1983, when Hershey moved on to the Naval Observatory. Coupled with Lippincott's retirement, the original Sproul observing program was, in effect, over. Heintz took the reins of the astrometry program, guiding it more toward the study of double stars—whether visual or actual binaries. Sproul began developing a spectroscopic instrument that could measure radial velocities—meant to be attached to the 61-cm refractor. The book was truly closing on Sproul's planetary research—but work was just ramping up in multiple other places.

1
2
3
4
5
6
7
8
9

"With one exception, the telescopes presently used in astrometry, that branch of astronomy concerned with the position and the motion of the stars, are about 60 years old, and only two were designed for high-precision astrometric observations."[1] This statement, from NASA's *Project Orion* proposal, kicked off an ambitious idea: update astrometry for the planetary era. The proposal, drawn up in 1980, brought together experts in the field, with Gatewood serving as an adviser.

Project Orion had no connection to the interstellar travel concept of the same name, which was a forerunner to Project Daedalus. (Rather than fusion, the Project Orion spacecraft would have just exploded a series of nuclear mini-bombs to get the spaceship going.) The NASA report recognized that as other areas of astronomy had advanced, astrometry had largely been left behind. The study hoped to modernize astrometry as well as find ways to otherwise detect exoplanets, through direct imaging or infrared signatures.

The ultimate project site would need to be geologically stable, with clear skies relatively free of distortion—the kind of weather that's the opposite of muggy Philadelphia. It also required room to expand and dark skies—making rural deserts the most promising places. The Black Mesa area of Arizona was considered the ideal site. It would have used automation and more stable instruments to "hold steady" when looking for planets astrometrically.

Despite a few feasibility studies, however, Project Orion never took off widely, perhaps because of anticipation around the upcoming space telescope, coupled with other emerging technologies that would supersede astrometry in planet hunting. Soon, the planetary science world would be revolutionized by a twist on the astrometric method: radial velocity.

New Beginnings

While the Sproul program was winding down, the hunt for planets was heating up, armed with new technology. Radial velocity had slowly crept into its own, and the astrometric hunt for planets had dwindled away. Allegheny Observatory, with George Gatewood's guidance, had continued an astrometric hunt for planets, updating its system with a new photometer, installed around 1982, to aid in the search.

A final report to NASA regarding the planetary search urged the importance of publication in obtaining funding while also taking a swipe at Sproul. "Most critical of our goals is to publish. It is through publications (as long as they are not about nonexistent planetary systems) that we gain the credibility needed in grant competition," Gatewood wrote.[2]

In the end, Allegheny's search yielded no concrete results in the 1980s. Ground-based astrometry had—and has, barring new breakthroughs—probably been stretched to its limits when it comes to searching for planets. The source signal is too small, the room for errors too large—a common problem in planetary detection, where a planet might be less than a percent the size of its star and even less than that the mass. Astrometry, however, remains a vital part of determining stellar positions, galactic distance, and the mass of two stars orbiting each other, even if planet hunters have pursued other methods.

Gatewood and his colleagues did suggest an interesting—and unutilized—solution for astrometry. While many in the astronomy community looked toward what became the Hubble Space Telescope for guidance in finding planets astrometrically, Gatewood and his coauthors suggested that an astrometric photometer (light sensor) could be placed on a US space station and yield positive results.[3]

Such a system would have had some unique advantages. Unlike Hubble, it could have been a facility dedicated to planet hunting and only planet hunting. Hubble requires shareholders across the astronomical community to request time on the telescope and use its bevy of instruments to perform separate investigations, but the proposed space station module could be used only for astrometry. Hubble hasn't been used much as a primary planet hunter, but it serves as one way to corroborate the detections of other agencies and watch known planets for things like atmospheric transits. The Sagittarius Window Eclipsing Extrasolar Planet Search (SWEEPS) program, which looked for transit events, is one case in which Hubble was used as a primary planet hunter, finding more than a dozen short-period worlds.

At the time Gatewood and his coauthors were presenting their paper, in 1987, NASA was, under the direction of President Ronald Reagan, working toward an orbital outpost for the United States. It would be the first US space station since Skylab fell in 1979. It would have been serviced by the space shuttle fleet and eventually be a launching pad to Mars. (That's what they all say, though.)

But Space Station Freedom, as it was more commonly called, never came to fruition. As often happens with NASA, it was the subject of cost overrun, and in the midst of the Clinton presidency (which was not, overall, the greatest time at NASA), the project was finally canceled. Instead, the work that had gone into Freedom went into a new venture: the International Space Station (ISS), which is still in orbit today and is run cooperatively by several space agencies. NASA, the post-Soviet Russian space agency Roscosmos, the European Space Agency, the Japan Aerospace Exploration Agency, and the Canadian Space Agency all share the station.

But while several of the goals of the US space station/Space Station Freedom have been fulfilled, ISS has never been used much for astronomical observation. The few telescopes that have been hoisted onboard have been used for Earth observations, perhaps because of the swift orbit, circling the world every 90 minutes, or the emphasis on science experiments surrounding long-term habitation in space. Whatever the reason, there's not a way to get a good close look at the stars on the 20-year-old (at the time of this writing) station.

Gatewood had struck on an idea for how to astrometrically search for planets from the ground, in what would be one of the last astrometry-based planet hunts from Earth. He and his team built the multichannel astrometric photometer, which involved a series of fiber optic cables hooked up to a computer. As is common with surveys today, a plate was created that selected a handful of stars, with a cable fed into the plate's face. Each plate acted as a sensor for that particular star in the night sky, detecting the faintest movements of stars and canceling out many of the noise problems faced by more traditional methods. What's more, by computerizing the efforts, most problems arising from badly exposed plates could be eliminated, helping shave down the margin of error experienced by the Sproul search.

The goal was to study a few dozen stars over the course of Jupiter's orbital period, 12 Earth years, hoping to see the kinds of perturbations a star might experience from a giant long-period planet in that span. Each session would take place over 45 minutes each observing night. It was a novel approach but limited in its applications, as it was "very delicate and time consuming for a very small number of stars to be widespread."[4]

The Allegheny program studied 15 targets in particular, but only found one of any interest: they detected a slight wobble around Barnard's Star. But it was written off as likely a statistical variation in an update on the matter in 1991.[5] A follow-up study didn't rule out the presence of planets but confirmed that any variation in the star had to be made by a planet of less than half the mass of Jupiter. This would rule out both of van de Kamp's planets.[6]

In fact, van de Kamp's planets have been whittled to almost nothing in astronomy. In 1999, the Hubble Space Telescope—that venerable machine—peered at Barnard's Star for an extended period. It found no evidence of short-period gas giants.[7] Radial velocity measurements—a mostly reliable way to find planets—found nothing in a 2003 study. It also ruled out not just close-in Jupiter-mass worlds, but ones similar in size to Neptune.[8] It seemed that most ground-based attempts at finding a planet via astrometry were doomed to fail, but perhaps a space-based mission really was the answer.

Back to Sproul

Swarthmore College worked, in the mid-1980s, mostly on the double star program, though one researcher worked (very) part time on finding the so-called Nemesis Star. In the mid-1980s, there were some rumblings about a possible small star or brown dwarf—called Nemesis—at the outskirts of our solar system, periodically bombarding Earth with comets and other debris. No such object was ever found.

John Gaustad continued to run the observatory, while Wulff Heintz continued to run the Astronomy Department. Rush Holt, who would later go on to serve New Jersey in the United States House of Representatives, worked at Swarthmore until 1988, when he moved to the State Department. NSF funding cuts devastated the college, forcing Borgman and others out of the department.

During this time, there was a little bit of planetary work at Sproul. For instance, their work debunked a planet at a time in the mid-1980s when it was suspected that the star ADS 11632 had a six Jupiter-mass planet. The Sproul program also ruled out one of its own candidates at Stein 2051. Around 1988–1990, Sproul researchers worked on several stars, trying to confirm several substellar candidates. This included Wolf 424, though that research pivoted to a new, more interesting scenario: that Wolf 424A or Wolf 424B was actually a brown dwarf. This idea was put forth by Heintz, who noted that their estimated masses—which he drove down from 7 percent the mass of the Sun to around 5 percent each—would make each too weak for nuclear burning.[9]

It was hailed as a "victory for old-fashioned astronomy" by the *New York Times*. The article also noticed that Wolf 424 had been Heintz's career obsession. That it was a binary system had been known since at least the 1940s, when Dirk Reuyl did some of the first in-depth studies of the star. But based on what Heintz could see through the 24-inch refractor, he wanted to confirm his suspicions that one or both were not true stars but instead the first ever discovered brown dwarfs.[10] It took the Hubble Space Telescope to finally resolve the masses, showing that both stars' masses had been grossly underestimated, and they were well above 10 percent the mass of the Sun . . . meaning they both remained very small stars.

Heintz also gave a provisionally substellar mass to GJ 1005B, an object he believed was 6 percent the mass of the Sun.[11] That, too, ended up being about twice the suspected mass. The binary system DT Virginis was also investigated by Sproul as a possible brown dwarf. The star versus brown dwarf debate for the smaller star isn't entirely settled—some estimates put it *right* at the limit for a star—so there's a provisional success here as well, and in 2010, a third component of around 11 Jupiter masses was found there.

But though the telescope was still base-level usable, it had several problems during this period. In 1987, the mirrors of the telescope had to be recoated because of environmental degradation, showing some of the encroaching creep of Philadelphia's urban pollution. The dome suffered damage in 1990, which took it out of commission for at least a month and a half, and the Grant machine was beginning to show its age.

The observatory's work was also split right down the middle now, between Heintz's double star work and Gaustad's work in infrared astronomy. Most of Gaustad's work focused on interstellar dust, with a little bit of work looking for *hot spots* in known binary star systems.

Otherwise, Sproul stayed largely out of the hunt for planets in the era of a more plentiful search for such objects.

Taking It Higher

Space-based astrometry was long a dream of the astronomical community because it didn't encounter distortions and errors introduced by the atmosphere. Using it for planet finding was discussed in earnest in the late 1970s, but some were still skeptical of its potential utility.

A 1983 paper by Ludwig Combrinck showed that space-based astrometry actually held great promise. Combrinck's paper still suggests that EV Lacertae, Epsilon Eridani, and Gliese 687 had giant "planet stars," a view that was a bit of an outlier at the time but perhaps shows how long it took for Sproul's observations to disappear completely from a planetary "canon." (In fact, the mass estimate of Epsilon Eridani's companion at six Jupiters would have placed it firmly as a planet. If it existed.) Of the failed detections of Barnard's Star planets, Combrinck said, "This does not reflect upon the quality of the work done at the Sproul Observatory, it merely indicates the difficulties encountered when attempting to detect extrasolar planets."[12] He suggested the use of space-based instrumentation eventually, while still finding potential in ground-based searches.

Gatewood continued to push for the space station attachment while describing other possibilities. At the 1986 AAS meeting, he and a group proposed a one-meter space-based telescope, believing that, from space, it could find planets down to the mass of Neptune.

The Soviets were working on Lomonossow, a space-based astrometry mission, in the 1980s. It was canceled when the Soviet Union fell, but elements of it survived and were eventually folded into ESA's Gaia mission.

Space-based astrometry was finally realized with Hipparcos, a European Space Agency mission to do space-based astrometry. The probe was launched in 1989 on a four-year mission to map the stars. It was the sort of space-based astrometry program that had long been discussed, but it encountered a few problems on the way up.

For one, the mission failed to achieve its target orbit. It was supposed to circle in a tilted orbit around Earth about 22,000 miles out. It achieved this at its farthest point, but a failed booster firing left it swinging within about 120 miles of Earth at its nearest approach, which also placed it in the cross hairs of Earth's radiation belts and atmospheric drag. While the technical definition of space begins at 62 miles, the Earth's atmosphere actually stretches quite a bit higher. The absolute-ish boundary of the exosphere is about 6,000 miles. That orbit was eventually pushed up to 300 or so miles, enough that it wouldn't affect the mission's lifespan.

This did hamper the craft's ability somewhat to serve as an all-sky survey, which may, in turn, have affected its planet-finding potential. (Although because of the sensitivity or lack thereof of the instruments, this possibility wasn't too seriously weighed.) While the mission proved essential in understanding the distance of several nearby stars, and discovering thousands of previously unseen stellar objects, it didn't yield anything new below that nebulous barrier between red and brown dwarf (or star and . . . not-star). Astronomers were able to eventually compare it to other planets discovered by other means, but the only compelling case it made for a planet were the observations of a few transits of HD 209458b in 1991. Had researchers been paying attention, they might have seen this odd behavior, but it went unnoticed until 1999. The 1991 observation is (maybe) the earliest direct evidence of a planet.

Gatewood intended to use Hipparcos data to confirm findings made by MAP. But even those were few and far between. A suspected planet around Rho Coronae Borealis was found to be a star in 2001, but MAP and Hipparcos otherwise overlapped very little (perhaps because MAP didn't have much success as a ground-based astrometry mission).

Looking for Heat Signatures

Some astronomers were beginning to search for planets in radial velocity signatures. Like astrometric planet hunting, it relies on looking for wobbles in stars. Unlike astrometric planet hunting, it doesn't look at the star directly but breaks down its light into constituent elements and looks for *shifts* in the light, indicating that something is moving the star toward or away from the observer.

But while radial velocity was coming en vogue in the small planet-hunting community of the 1980s—estimated to be a few dozen deep—others were beginning to search for planets in infrared. Heat, like that given off from the Sun, drives up the temperature of objects near it. This, in turn, glows in a red light just a few steps above what our eyes can see. This is called *infrared* and is a good way to detect objects that don't necessarily give off light of their own.

In 1983, NASA launched the Infrared Astronomical Satellite (IRAS), a testbed for infrared astronomy from space. IRAS performed a sky survey, discovering several asteroids and other objects in our own solar system that give off little light. But IRAS was fairly primitive, meant mostly as a testbed for space-based infrared observations, and in 1983 it gave off a false planetary detection—this one much closer to home.

"A heavenly body possibly as large as the giant planet Jupiter and possibly so close to Earth that it would be part of this solar system has been found in the direction of the constellation Orion by an orbiting telescope aboard the US infrared astronomical satellite," the *Washington Post* reported. "So mysterious is the object that astronomers do not know if it is a planet, a giant comet, a nearby 'protostar' that never got hot enough to become a star, a distant galaxy so young that it is still in the process of forming its first stars or a galaxy so shrouded in dust that none of the light cast by its stars ever gets through."[13]

This, in turn, fed into a brewing hypothesis of a Nemesis Star, an object at the outskirts of our solar system that could be the reason for mass extinctions on Earth. It was discussed as a possible high-mass brown dwarf or low-mass red dwarf. But search after search failed to find the object (though some in the absolute fringes of pseudoscience swear it's there as an astrological object called Nibiru, pushed by new agers and UFO enthusiasts).

But IRAS never discovered Nibiru or Nemesis, and it never discovered a "tenth planet." (Let's not wade into the Pluto debate right now.) The object in question—and three other mysterious objects—ended up being distant, fuzzy galaxies. It was probably the more realistic conjecture, but a new galaxy—one of 2 trillion or so—is less likely to cause a stir in the press than a new planet in our own solar system.

IRAS did make a few important planetary-adjacent discoveries though. In 1984, it discovered the first ever debris disk around another star, Vega, a star that is very young, very bright, and, at 27 light years away, fairly nearby. A debris disk is an asteroid-belt-like ring or belt of materials in orbit around a star that consists of the building blocks of planets. While not a planet in and of itself, it can point to the distinct possibility of planets, such as if a path is cut through the belt.

IRAS didn't directly see the belt. Instead, it saw the presence of infrared excess, which is when a star heats up the dust around it, giving off a faint glow in infrared telescopes. It was the first incontrovertible proof of planetary formation, though to date no planet has been found in the system (though there is evidence that there may well be planets that are hard to detect because of the alignment of Vega's planetary plane toward Earth). Vega is so young, however, that it may even be actively forming planets now. Most studies have ruled out anything larger than Jupiter, but smaller objects may be forming in the dust. But even if it doesn't mean planets per se, the presence of a debris disk points to the formation of comets, asteroids, meteors, and other natural space debris.

The debris disk around Vega had a famous appearance in pop culture. Carl Sagan used it in his only novel, *Contact,* as the place for a wormhole that transports the novel's heroine, Eleanor Arroway, off to a distant planet. In the novel, aliens send a message to Earth through a relay satellite in the vicinity of Vega. This was altered slightly for the 1997 movie adaptation of the novel, starring Jodie Foster, though the location of Vega remained the same. The big blue star is contrasted against the debris disk in the movie as Foster is transported to the system, only for the film's brilliant SETI scientist to plunge into another wormhole and see an alien telepathically appearing as her dead father.

Sagan was a perceptive scientist who worked toward bringing science into the public imagination through his books and the PBS series *Cosmos.* But from the beginning of his career, he also was fascinated by the possibility of life beyond Earth, some of it intelligent. His first book was *Intelligent Life in the Universe.* As belief in van de Kamp's planets waned, Sagan still mentioned them in *Murmurs of Earth,* a book he coathored with Frank Drake, Ann Druyan, Timothy Ferris, Jon Lomberg, and Linda Salzman Sagan, about their creation of the Golden Record, an object affixed aboard the Voyager crafts, both bound out of our solar system to parts unknown, both returning the faintest murmurs of the outer solar system back to Earth today. "Our ability to detect planetary systems around other stars is at present extremely limited, although it is rapidly improving," they wrote. "Some preliminary evidence suggests that there are one or more planets of about the mass of Jupiter and Saturn orbiting Barnard's star, and general theoretical considerations suggest that planets ought to be frequent complements of most such stars."[14] (The latter point has, indeed, proved more than correct in the intervening time.)

Sagan, until his death in late 1996, kept abreast of each and every discovery regarding planets and SETI, so perhaps it's no surprise that even something so new as Vega's protoplanetary disks would appear in his works as the most recent discovery of planets. Maybe in a different world, one where Gatewood had never intervened, Arroway arrives in Barnard's Star system, only to be swiftly whisked away—though had Gatewood not caught it, someone else likely would have by 1986.

Similarly, Fomalhaut and Beta Pictoris were found by IRAS to have protoplanetary disks. Fomalhaut is part of a three-star system and has an extremely diffuse debris disk affected by this system, while at least one gas giant sweeps through. Instead of shaping the belts, the orbit of Fomalhaut b—popularly called Dagon after a public vote—seems like it should be ripping it to shreds. The two distant companion stars may be what controls this chaos.[15]

Beta Pictoris is similarly known today to have a giant planet. It has three debris rings, with evidence for three more planets possibly hiding somewhere in the system. Both Beta Pictoris b and Dagon hold important distinctions: they are two of only a few dozen planets to be spotted directly, as small smudgy pixels in the midst of their rings. Directly imaged planets are always young and still hot from formation, which is why they appear in infrared with a little finessing and a little occulting (blocking out the light of a star in order to find such objects). The planets may even have giant Saturn-like rings. There's also some evidence that Beta Pictoris b could have a moon. If confirmed, it would be the first exomoon discovered. A Columbia University team is currently trying to confirm an exoplanet candidate around Kepler-1625 b. That has a unique configuration: the parent planet is the size of Jupiter, while the "moon" candidate appears to be the size of Neptune![16]

But in 1983, all that IRAS could see were belts appearing like jets emanating from Beta Pictoris, a messy smattering of dust that ended up being two debris disks, one inclined from the other. At that point, no infrared telescope could have seen the young planet creeping slowly through the system, completing an orbit every 7,890 days. It similarly couldn't have seen Dagon, a lumbering beast that takes 1,700 *years* to complete just one orbit.

There is an interesting fold to the Beta Pictoris study, though. Beta Pictoris b was officially announced in 2008 via direct imaging—that is, it was seen in infrared as a distinct object. As early as 1993, there had been some evidence of comets or other small rocky bodies in the system via spectroscopic measurements. But in 1981, two years before IRAS, astronomers witnessed a brief dip in the light curve of Beta Pictoris. No one knew quite what to make of it—a giant comet was suspected—but after the official discovery of the first exoplanets, there was a little bit of chatter about it possibly being a planet, though no conclusions were officially drawn. The initial observations of the planet in 2004—leading up to the 2008 announcement—gave a period of a little less than 19 years. The opportunity to spot a transit in 2000 had passed—indeed, photometers weren't quite at the right technological levels at the time—but astronomers believe the planet, if it does transit, transited again in 2018. If this proves true, then it's possible that Beta Pictoris b, which is widely accepted by the astronomical community as a planet, was the culprit behind the 1981 transit. This would make it the first detection of an exoplanet.[17] Some studies after the transit window, however, have said that it's not likely to have been the culprit.[18]

These three discoveries in 1983 were the first of more than 100 debris disks IRAS would go on to discover. Stars were thus confirmed to form dust—which provided the necessary material to coalesce into rocks, larger rocks, really large rocks, dwarf planets, protoplanets, and planets. But actual planets seemed to evade detection. There were eventually detections of brown dwarfs in infrared in 1988—Gliese 569B and GD 165B. Astronomers weren't entirely sure what the objects were at first, though. They were low temperature and very dim, seemingly constituting a new class of object. Astronomers just weren't sure what that object was. They called it an L-dwarf at the time, with a caveat: "While the classification as L-type is indicative of a rather low atmospheric temperature (below approx. 2,000 K), it is not a guarantee, however, for the absence of hydrogen-burning in the interior."[19] Astronomers had to prove that lithium—a heavier element that burns up readily in most stars—was abundant to disprove they were stars. And around this time, a whole lot of candidate objects were straddling that same boundary between star and brown dwarf and had yet to be confirmed one way or the other. Like the search for planets—a search to which brown dwarfs are forever tied—the search for brown dwarfs was full of stops, starts, tentative detections, ambiguous objects, and other problems that made crossing the finish line into a "first" discovery difficult.

In 1984, astronomers at the University of Arizona, using the Steward Observatory, believed they'd hit the planetary jackpot. Or maybe it was the MACHO jackpot. The star van Biesbroeck 8 (vB8), sometimes called V1054 Ophiuchi C, is one of five stars in the V1054 Ophiuchi system, which is around 21 light years distant. It's the smallest star in the system, just 8 percent the mass of the Sun, making it just barely a star.

Because the system is nearby, it's not hard to dissect, despite its seeming chaos. The faint stars burn in infrared, including vB8. But an infrared look at vB8 appeared to show something else, something roughly the size of Jupiter. It gave off only a faint amount of light and didn't seem to be a star, but based on tracing the movements of vB8 astrometrically, it seemed to be too massive to be a planet. It was, at a minimum, 30 Jupiter masses.

In 1984, there were no confirmed planets, but here, the first brown dwarf seemed to be coming into view. It was the holy grail of low-mass companion research in the 1980s, because astronomers were puzzled as to why they hadn't discovered such an object yet.

The not-planet, not-star was first predicted in 1983.[20] Both vB8 and another low-mass star, vB10, seemed to be moving *oddly,* in astrometric terms, but it wasn't enough to be a star. The infrared trace seemed to be a confirmation that vB8 had a verified brown dwarf. Gatewood called it a failed star orbiting a true star in an interview with the *New York Times*.[21]

The brown dwarf versus planet debate ramped up.

But the story quickly fell apart. In short, vB8 never reappeared. William McCarthy, one of the discoverers of vB8, told the *New York Times* that he hadn't been able to reobserve the brown dwarf. "Dr. McCarthy said he could conclude only that this meant the original observations were wrong or the orbiting companion object had now moved in closer to the star and the two could not be distinguished from each other," John Noble Wilford wrote in 1986.[22]

To date, vB8 has never been recovered and was likely an error of astrometry and a chance background object. Researchers pressed on, searching for the next substellar object. They thought they had it in a brown dwarf around the degenerate star G 29-38, but that never transpired, either. Instead, the spotted object was a clumpy part of the star's debris disk. "The astronomers said there was only 'one chance in a million' that they were detecting radiations from another star passing in the background of Giclas 29-38," Wilford wrote in 1987. "They said it was unlikely that a swarm of asteroids or dust grains surrounding the star could account for the radiation, for it is difficult to imagine how they could have survived the cataclysm that Giclas went through more than 600 million years ago."[23]

But the objects—likely comets—did survive. And they certainly didn't have the mass to measure up to a brown dwarf. It was another misstep in the gradual discovery of planets outside our solar system, and not brown dwarfs at all.

There are two small footnotes to the search for astrometric planets, which will take us briefly out of the timeline in the 1980s and into the last 10 or so years. One purported astrometrically discovered planet is around a star called vB10, which held the record as the faintest known star from 1944 until 1983. It orbits the comparatively brighter star Gliese 752. In 2009, astronomers put forth the idea that a six Jupiter-mass planet orbited the star.

The discovery of vB10b is contentious. Jet Propulsion Laboratory (JPL) astronomers using the Hale telescope at Palomar claimed to find the planet *astrometrically*. But radial velocity measurements didn't find anything there. A team led by Jacob Bean, of the University of Chicago, didn't mince words in their paper, "The Proposed Giant Planet Around VB 10 Does Not Exist."[24] The paper found that a planet any larger than three Jupiter masses couldn't be there. This put a significant ding in the work of the JPL astronomers, who thought they, at last, had the first ever confirmed planet discovered by astrometry. But it was not to be.

There's also an incredibly tentative detection of the planet HD 176051b. It was detected astrometrically in 2010, but no one knows quite sure how it orbits the binary pair of stars in the system or, indeed, much about it, and it's still considered a candidate planet, rather than confirmed. If follow-up data confirms the presence of the planet, it would be the first planet found by astrometry.

The last, best hope for a solid astrometric planet comes in the form of the Gaia mission, ESA's successful follow-on to Hipparcos. The spacecraft is conducting an all-sky survey that has taken in an astounding 73 terabytes of data about our galaxy and other nearby galaxies. With an expected lifespan into the 2020s, that could become petabytes. From these data, researchers expect anywhere between 21,000 and 70,000 Jupiter-like planets could be found through astrometry.[25] But so far, the data have yielded no planets.

Any day now. . . .

At the very least, there have been a handful of larger brown dwarfs found through astrometry. The brown dwarf candidate DENIS-P J082303.1-491201 was found astrometrically, weighing in at a hefty 28 Jupiter mass. Some catalogs consider it a planet, though this could be the result of not having an exact mass on the primary "star," which is itself believed to be a brown dwarf. The brown dwarf GJ 802b was first reported in 2005 and has a mass just a hair below 60 Jupiter masses. And the mysterious maybe-star LSR1610-0040 has a definite astrometric disturbance that's either a very small star or a very large brown dwarf.

The Future from Above

In 1986, the NASA advisory council released a report regarding the far future of the agency with the new millennium approaching. There were calls for a Mars sample return mission—which is on the cusp of possibly happening with the Mars 2020 rover—as well as asteroid mining, comet sampling, exploring the gas giants, and more. It also made a recommendation for NASA to install the space station–based observatory.

"Theories of solar system formation have become advanced enough to predict the general existence of planets, and ground-based observations, following up the first space infrared sky survey, have discovered what may be a planetary system in the early stages of formation," the report states.[26] The IRAS discoveries may not have found planets, but they were certainly hastening the search to find them.

The report further notes, "The solar system is not, in face, a unique or singular object"—the drum van de Kamp had beaten for years. A strain of thought that often crops up in the absence of a discovery is that we have to contend with possibly being the only place with conditions like ours in the whole universe. For instance, because SETI's 60-year search hasn't yielded anything yet, some believe in the rare earth hypothesis, that our planet may be the only place with all the conditions to bring up intelligent life from simple microorganisms. While this hasn't been destroyed entirely as a line of thought, planetary discoveries in the interim have ever so slowly chipped away at it.

There are even rare solar system hypotheses today, that our planetary arrangement is hard to come by, but most of our data come from a few years, rather than several decades, of study. If a world with a Saturn-like period transits its star from our perspective, we would see a transit event only every 29 years or so. To capture the three-ish transits necessary to get a confirmation, we would have to study that system for *87 years*. Compare that to a planet with a Mercury-like 88-day period, which could witness four transits in a hair under a year. Beliefs about the likelihood of certain findings often rest on an observational bias.

There was even the thought, at some point in the 1980s, that brown dwarfs might not exist, simply because we'd found none to date. Planetary detections had been just as elusive (unless you asked Peter van de Kamp), but Vega's debris disk made the search seem less nebulous to some, who realized that even though our solar system may have its own unique qualities, it likely wasn't alone. In 1917, the Carnegie Institution for Science captured evidence of a debris disk at Van Maanen 2 in spectroscopic measurements, but it wasn't realized until 100 years later—this could have eclipsed the Vega discovery by 70 years had anyone thought to look in the intervening time.[27] But these moments also serve as a kick in the pants to researchers to push observations further. We may have found several dozen spectroscopic signatures in the intervening decades, but technology still has to come into its own to make the discoveries possible. In the 1930s—before Strand's erroneous discovery—astronomers like James Jeans and Arthur Compton believed that planets were rare, that chance encounters between stars created them. Better models and more evidence refuted that.

Vega was, in some ways then, the catalyst for our modern pursuit of planets—the first to bear fruit. The solar system was, officially now, not a singular and unique phenomenon. The NASA group thus had a few ideas for how to search for them: astrometry, radial velocity, the transit method (called "luminosity measurements" in the Solar System Exploration Committee report), direct imaging, infrared hunts, and interferometry—using several observatories to simultaneously observe the same object and derive objects from the data. Of those, all but astrometry and interferometry have paid off. A NASA program called SIM—Space Interferometry Mission—was repeatedly slated for launch before being canceled in 2010. It would have used two onboard telescopes to make subtle astrometric observations and find objects by comparing observations between the two. Gatewood, prior to SIM, had pushed for the Astrometric Imaging Telescope, another space telescope based on using astrometry.

"The abundance of planetary systems in the universe is a fundamental piece of objective data which we must have in order to construct, test, and modify our theories," the report states. "The discovery of other planetary systems, and the study of them at a level sufficient to reveal structural regularities like those which characterize our own system, have now become essential if we are to turn our theories about star and planet formation into a real and objectively based science." It goes on to say that "from the technical point of view, the capabilities needed to search for other planetary systems are also coming within our reach"—and that the biggest barrier had been the atmosphere, which is exactly why the report suggests hoisting everything up into orbit. The report authors found that such efforts—especially using the space station as a platform—were essential in finding whether, at the very least, our solar system was rare, or even alone:

> An extensive search for other planetary systems, whatever the outcome, cannot help but have a major impact. If other planetary systems are found, then new directions will open up in both science and philosophy. Our own planetary system will become firmly established as only one specific instance of a general phenomenon.
>
> On the other hand, if no other planetary systems are found after an exhaustive search, we will have to admit that planetary systems like ours are uncommon, and perhaps unique, in the universe. Our ideas about the formation of our solar system will have to be reexamined in the light of evidence that the system is a cosmic accident, some freak of nature, and not representative of anything general.[28]

The report authors believe that MAP serves more as a proof of concept for a space station astrometry module, as "rigorous detection and study of planetary system" could never achieve the right accuracy from the ground. Astrometry, by all accounts, was far out of favor as a planet-finding method.

Blink of an Eye

The transit method (sometimes called *photometry*, based on the instrument used) is perhaps one of the most interesting approaches to finding planets. It measures the amount of light coming from a star and looks for miniscule dips in that light at regular intervals. From the depth of those blots, it can determine the size of an object passing in front of it from our point of view and show whether or not it's a planet. It can be done from the ground with a small telescope, easily finds planets by dips, and can even survey several stars at once—as demonstrated by NASA's Kepler mission.

It was common knowledge that finding a planet via transit was possible, but that instrument sensitivity wasn't quite there yet into the 1980s and even the 1990s. As photometers became better and better—driven in part by advances in digital imaging—the technique seemed right over the horizon. Indeed, several studies in the 1980s indicated the potential for the method. (And if Beta Pictoris b was, in fact, detected in 1981, the method had already paid off.) While radial velocity is good for finding very large planets, it is not especially great at finding very small ones. The transit method requires finding a dip in light when a planet crosses in front of a star and doesn't rely on mass to make observations.

A 1985 NASA Ames paper outlined the abilities of photometry in finding extrasolar planets, suggesting that it could even be used to monitor stellar phenomena like flares, star spots, and other brightening and dimming events. This is partly because photometry is not an especially elegant method. It relies less on the telescope and more on the sensor. That may be part of why, despite this initial report suggesting that it could be a potential way to find planets of varying sizes and masses, the first planet detected by transit didn't happen until 2002, and efforts at detection didn't begin in earnest until the end of that decade. Sensor technology just wasn't there yet.

The report made an interesting proposal: the transit method could find planets en masse by measuring up to 1,000,000 stars in a field. In 2009, NASA launched the Kepler telescope, which stared in the direction of Cygnus, monitoring 150,000 stars at once. To date, it has found 2,300+ confirmed planets and identified another 4,000+ candidates awaiting confirmation.[29]

Thus, the potential of transit was well recognized in the 1980s, but the key component in it all—a sufficiently sensitive sensor—required a CCD (couple-charged device) advanced enough to capture small dips in light. That took advances in digital photography not entirely possible until the late 1990s.

One idea that for sure wasn't happening—especially at the time it was proposed—was to move beyond Earth in hunting for planets, but not in the same way as a space telescope does. The proposal, put forth by Bernard Burke, sought to put an observatory on the moon, consisting of an infrared telescope, an interferometry array, and a large optical telescope with an aperture of 16 meters. That's twice the size of anything online as of March 2018, though a little more than half the size of the planned Thirty Meter Telescope. Burke clearly wasn't thinking short term with the proposal, saying that the observatory "probably will require assembly and maintenance by lunar-based personnel, or by robots controlled by those personnel through a tight servo-loop." Thus, it was an interesting idea, but one that's unfeasible in 2020, let alone 1990.[30]

Astrometry was out of the question, and the first interferometric candidate—vB8—was out of the running. Infrared methods were sensitive enough to find massive structures like debris disks, but not to suss out individual objects—at least not until 2004. This put exoplanet researchers back at radial velocity as the only viable option.

Gamma Cephei

In 1988, astronomers Bruce Campbell, Gordon Walker, and Stephenson Yang believed they'd finally, finally, finally found a planet. This was as van de Kamp was making his final pleas for the existence of his own worlds. Radial velocity measurements of Gamma Cephei indicated a low-mass star in orbit around the primary star, which is larger than the Sun. Gamma Cephei A is beginning its death spiral, bulging out as an orange subgiant, while its smaller companion, Gamma Cephei B, will burn on for trillions of years because it's a red dwarf star about Uranus's distance away from its primary.

Because Gamma Cephei A is large and bright, even a more massive red dwarf like Gamma Cephei B hides easily from optical telescopes—but not from radial velocity. As Gamma Cephei B tugged on on its larger sibling Gamma Cephei A ever so slightly, it disrupted the light in a faint ripple or wave. The ripple was big, clearly a star. But there was a smaller ripple, too, that seemed separate from Gamma Cephei B, having a less significant tug—something around two Jupiter masses. Gamma Cephei Ab seemed to be the first exoplanet ever officially discovered, though astronomers still weren't entirely convinced.

The paper that first introduced Gamma Cephei Ab as a strong planet candidate included a few interesting possibilities, including a potential (weak) candidate at 61 Cygni B.[31] Looking back at Strand's early Sproul work, the authors also suggested that the Xi Boötis system—which Strand suspected had a companion but he never delved deeply into it—could indeed have a planet. There was also evidence in the paper for something much closer to Jupiter, mass wise, at Epsilon Eridani. (There's a promising—though unconfirmed—planet right around that mass in the system.) Another star reported to have a companion in the Sproul program, 36 Ursae Majoris, was also tentatively identified by Campbell, Walker, and Yang as having evidence for an object in its orbit. Beta Virginis, 61 Virginis (not to be confused with 61 Cygni), and Beta Aquilae A are listed as well. Of those three, only 61 Virginis has confirmed planets thus far.

This was the same case with HD 114762b, which is significantly heftier at 11 Jupiter masses. HD 114762A is quite Sunlike, while HD 114762B is more of a midrange M-dwarf. HD 114762B completes an orbit in 84 days, stretched out into an eccentric orbit by the interactions between A and B.

Campbell was, in many ways, the reason that radial velocity was coming into its own. Then a grad student at the University of British Columbia, he worked with Walker on figuring out a way to use the *Doppler method* to find unseen objects. The Doppler method studies subtle shifts in the velocity of a star caused by an unseen object tugging on it, causing its light spectra to become either bluer or redder from the perspective of the observer.

It would just be really, really hard to use Doppler while trying to find small variations in the rotation of a star to figure out if something is tugging on it. It's not going to be super apparent by just staring at the star with a conventional telescope. But if you look at one with a spectrograph—which monitors the light output of a star to determine its constituent elements—you might be able to make subtle deductions about the presence of other objects. The method was first proposed in 1952 by Otto Struve, and astronomers deduced that Jupiter would cause the Sun to wobble at about 12.4 meters per second over the course of its orbit. But early spectrographs were incapable of finding a signal that subtle. It was thus largely used to study binary stars in efforts to determine the mass of one of the stars in the pair.

Campbell pioneered a way. He installed a cell that would use trapped hydrogen fluoride gas to filter the light of a star. Hydrogen fluoride is nasty, nasty stuff. One of its main industrial uses is in hydrofluoric acid. This also makes it quite corrosive, so it would wear away at instruments, including spectrographs. But the hydrogen fluoride cells, before eating away *too* much of the glass, could drive the detectable speeds down to 13 meters per second—close to being able to find a Jupiter analog around another star. "The starlight is collected by the primary mirror and passes through the *gas absorption cell* just prior to entering the spectrometer," wrote exoplanet pioneer Paul Butler for Carnegie Science. "The spectrum of the gas vapor in the absorption cell is imprinted on the starlight and provides a reference spectrum against which to measure the Doppler shift of the star."[32] Butler would later pioneer a much, much safer method. The light needed to pass through a colored gas in order to pick up the spectral clues. Butler was able to eventually create a gas cell filled with iodine—less dangerous and corrosive by far.

Gamma Cephei Ab strained and struggled to get its planetary status certified. By 1992, the astronomical community had decided that the radial velocity signature was too weak for there to be a 2 Jupiter-mass planet there. After an initial flurry of activity, it seemed to recedeinto the background, with even the original researchers admitting that the little ripple seen may have been just an artifact or some other activity around the star.

Meanwhile, HD 114762b was confirmed in 1991, but nobody was sure what they were confirming. The signal showed a minimum mass of 11 Jupiter masses—which would strain the line between brown dwarf and gas giant—and a maximum mass that could have strained the line between brown dwarf and red dwarf. In other words, no one was quite sure what it was. The initial paper called it a "probable brown dwarf," admitting that it was hard to nail down any details about it.[33]

That was a difficult delineation to make at the time, as more and more seemingly substellar objects were discovered, but no one was quite sure what they were looking at. As *Astronomy* magazine ran the article "Do brown dwarfs really exist?" in 1989, a group of researchers announced 10 candidate brown dwarfs in the T Tauri association.[34]

Most times, astronomers debated if the objects they captured in infrared were red or brown dwarfs, or if a spectroscopic companion was a star or a failed star. HD 114762b was perhaps the first time—though certainly not the last—that anyone invoked the "brown dwarf or planet" debate on an object confirmed to exist. Some astronomers, though, find that a high-mass planet and a low-mass brown dwarf may not be so different after all. This is especially true today. "Some of this new work hints that the brown dwarfs and planets may not be distinct types of objects," noted Ben R. Oppenheimer. "That they may be intrinsically related, or that the two categories have significant overlap in properties. Time will tell, and perhaps some of the unraveling of the mysterious connection between stars, brown dwarfs and planets will come from the new ability of astronomers to conduct what I like to call 'remote reconnaissance' of exosolar systems."[35]

But because HD 114762b was largely believed to be a brown dwarf, the 1980s closed out with only one planetary candidate—Gamma Cephei Ab, which would be ruled out within just a few years.

Nonetheless, the 1980s would be the last decade without a confirmed exoplanet. The 1990s finally saw the true dawn of planetary discovery—and of the promise held in searching for a star's wobble. It just ended up that van de Kamp and crew were looking for the wrong kind of wobble.

1
2
3
4
5
6
7
8
9

In 1992, astronomers discovered the first planet outside our solar system—and this time they were serious. But it didn't come in any form they'd really anticipated.

Neutron stars are the second densest type of object in the universe outside black holes. When a giant star dies, it explodes outward as a result of the collapse of its core. Put simply, the star becomes too massive to go on and expels all its energy into the surrounding space. The core is a sort of ground zero of this detonation. When that core collapses, depending on the size of the star, it becomes either a neutron star or a black hole.

Some neutron stars are called *pulsars*, for the regular "pulses" they give off in radio frequencies. Think of many of them like a drummer—fast regular beats. Some pulsars, called *millisecond pulsars*, "drum" so fast that it would put Napalm Death's drummer Danny Herrera to shame. And much like the grindcore played by bands like Napalm Death, these pulsars have an internal fury—the radio pulses come from the rotation at the core of these objects, which are the size of a small city but the mass of an entire star. (Of course, the pulsar RCW 103 pulses only once every six hours, so not every pulsar is cut out to play fast-and-furious heavy metal.)

Those pulses are so regular that if they don't come at the right interval, astronomers know something is off. For instance, astronomers detected that something was off in the Crab Pulsar when its metronome wasn't quite timing itself correctly. They excitedly thought that perhaps a planet was present, but it quickly became apparent that the interference was caused by the nebula gas around the supernova remnant.

There's also a 1979 case of a purported planetary system at PSR B0329+54. It was reported to have two planets in orbit around it, which would have made a spectacular first announcement.[1] A 1995 study seemed to confirm it.[2] A few subsequent studies echoed these findings, but just two years later, those discoveries all but evaporated.[3] In late 2017, though, another study came along stating that maybe planets were there after all.[4] It is . . . quite the back and forth, a debate that stretches 40 years without a true answer. (If finally, finally, finally verified, they would be the first planets detected outside our solar system, though not around a normal star.)

A breakthrough in 1992 provided rock-solid evidence of planets with a near bulletproof case. Astronomers Aleksander Wolszczan and Dale Frail tuned into the pulsar PSR B1257+12, 2300 light years away.[5] It should have pulsed every 0.006219 seconds, but every now and then, its pulses were a little off. Yet those off-beats came at regular intervals as well. After intensive study, Wolszczan and Frail came up with an explanation for why that was: it had two planets around it. One was three and the other four times the mass of Earth, and they rotated around every 67 and 98 days, rounded up.

Finding planets around pulsars is very, very hard, as Andrew Lynne of the University of Manchester had figured out just a year earlier. A proposed 1.4 Earth-mass planet around PSR B1829-10 was actually a math error. Lynne hadn't accounted for the rotation of Earth in the measurements initially, and once it was factored in, the irregularity disappeared.

In fact, since 1992, only two more pulsar planets have been confirmed. In 1994, a third planet was discovered in the B1257 system, this one less massive than our Moon—making it still the least massive exoplanet known. The fourth came in 2003 with the discovery of a gas giant around PSR B1620-26. It's a 2 Jupiter-mass world orbiting the center of mass between a neutron star and a white dwarf—two dead zombie stars.

Pulsar planets are somewhere in between a zombie and a chimera. When they explode, usually the planets in that system are destroyed or flung out by a shockwave. But after the violence settles down, the gas and dust can recondense. This, in effect, means that the three planets in B1257 may be made out of parts of the planets that came before them.

Given the extreme radiation in these systems, almost no one has ever thought that the B1257 system could host life. A 2017 study seems to be an outlier. Astronomers Alessandro Patruno and Mihkel Kama ran simulations on the system and discovered that if all the right factors came into place, and the planets were the right size, they could get a gas envelope and enough of a chemical soup to become habitable.[6] This scenario isn't likely, however.

So astronomers thus had the first verified planets around another star, but no proof of planets around a *main sequence* star like the Sun.

That kind of confirmation was still a few years away.

The Groundwork

When Lippincott retired in 1983, she ceased publishing. This was in contrast to van de Kamp, who seemingly continued on into the mid-1980s, publishing *Dark Companions of Stars* in 1986, more than a decade after he had nominally retired. Both were fairly quiet in the 1990s, as interest in finding planets continued to heat up. Hershey continued to work on the space telescope program. His Hubble work interrogated several low-mass binary objects, looking for brown dwarfs, but instead turning up faint low-mass stars. He assisted in the discovery of the previously mentioned—and unconfirmed—object around Proxima Centauri. He was also part of a team that confirmed transits around a planet previously discovered by radial velocity, HD 209458b, in 2001. He split his work at Hubble between stellar studies (usually looking for brown dwarfs or planets) and studying the Neptune system.

Though Hershey did work with data of real live planets, he never found a fully confirmed one—as is the case with Lippincott. Both Hershey and Lippincott found a few objects that haven't been fully erased from the pantheon of planets, unlike Peter van de Kamp, whose *planetary* work was largely stricken from the annals of astronomy aside from footnotes and margins.

But van de Kamp, for as much as he became synonymous with Barnard's Star, left behind a spectacular body of work when it came to the positions of the stars nearest Earth, and with discoveries of hard-to-find stars, obscured by a brighter companion or otherwise hidden from our view, but not from the exacting eye of van de Kamp or the Sproul Observatory. After all, the flaws in the telescope only crept in when looking for very small objects, those smaller than a star—but it worked just fine for finding some of the smallest stars ever discovered.

Lippincott's name is more associated with Ross 614B than Lalande 21185 in a way that Peter van de Kamp's, unfortunately, is not.

Geoff Marcy and Paul Butler worked on detecting exoplanets from the 1980s on. In the book *Other Worlds,* author Michael D. Lemonick follows Butler and Marcy's quest. He writes that they had a fear "that they'd someday be mentioned in the same sentence with Peter van de Kamp."[7] (After allegations of sexual harassment surrounded Geoff Marcy and forced him out of the University of California,Berkeley, there's a lot more reason to fear being mentioned in the same sentence as Geoff Marcy today.)

Gatewood relays a similar story in *Other Worlds*. Every time he thought he had found a planetary system, he would reach out to David Black at NASA after heavily scrutinizing the claim. "George, are you *sure?* You don't want to become known as the Peter van de Kamp of your generation," Black would tell Gatewood.

In addition, van de Kamp's books, like *Principles of Astrometry, Elements of Astromechanics,* and *Basic Astronomy*, are landmark books in the field, still used by practitioners of astrometry in the search for dark companions. Van de Kamp had also correctly deduced that there was validity in looking for planets via a wobble in a star's movements, but his earnest pursuit for such objects ran up against the limitations of ground-based technologies.

His ardor in proving his most famous work—that around Barnard's Star—perhaps also overshadows the general respect he had from peers and his popularity among colleagues and students. Once a fixture of astronomy, his relative quiet from the 1980s on left him receding from the popular imagination but didn't diminish his contributions to our understanding of the position of stars. Astronomer Doug Braun once said, "Van de Kamp pretty much defined the field of long focus astrometry, and his contribution to astronomy is profound," in rebuttal to a colleague misstating the nature of the Barnard's Star affair.[8]

On May 18, 1995, van de Kamp passed away. He was 93. Lippincott wrote his obituary for the American Astronomical Society, calling him an "authority in the field" of astrometry, and highlighting not only his astronomical career, but also the well-rounded, friendly, gregarious devourer of Charlie Chaplin films and classical music, whose dedication to the latter had given him a special composition from composer Peter Schickele, the sort of Weird Al of classical music who went by the name P. D. Q. Bach. The piece was called "The Easy Goin' PVDK Ever Loving Rag." Lippincott wrote of her mentor that he "always expressed the belief that Astronomy was a marvelous synthesis of art and science, and he patterned his successful life in that fashion."[9] David Stout wrote van de Kamp's obituary for the *New York Times*, stating that as far as Barnard's Star was concerned, "His calculations had been right; the wobbling, alas, was not in the star but in the telescope," and reflected on the humanism van de Kamp brought to the field: "Dr. van de Kamp was a musician and composer who directed the Swarthmore College Orchestra from 1944 to 1954. 'I do not see any conflict between humanities, art and science,' he said in a 1972 talk. 'I consider science just another liberal art.'"[10]

Laurence W. Fredrick, memorializing his friend in *Publications of the Astronomical Society of the Pacific*, pointed out that van de Kamp's earliest work came in using astrometry to determine that the distance to the center of our galaxy was 12,000 parsecs, or 39,000+ light years, a figure that largely stayed in place until the Hubble era. Fredrick wrote, "Van de Kamp was very conservative in his approach to science: the integrity of his work could not be questioned. . . . When a new approach appeared promising, viz., digital computers, he made reductions in parallel until he was satisfied the new approach worked, the same with measuring machines." Fredrick also shares an anecdote: "One time when I pressed van de Kamp to invest in an autoguider he declined on the grounds that he could not see what the electrons were doing. Actually, at that time, autoguiders had not been reliable and there was no point in spending money on an unreliable instrument."[11]

He also regaled with stories of van de Kamp the home entertainer. If a professor came to visit, Fredrick said, van de Kamp made it a learning opportunity for his students—and would often have an informal party, replete with German beer and a more casual atmosphere than the stuffy confines of Swarthmore, which was steeped in the temperance trappings of Quakerism. There were also little details that escape a lot of other details of van de Kamp's life, like an amateur film-making hobby, his own musical compositions, or the time that, according to Fredrick, van de Kamp cut out of an IAU meeting in Brighton, England, to go to a Buster Keaton film festival. He ended up playing piano music for a bit after an intermission. Also, in 1938, van de Kamp played in a home quartet with Albert Einstein on violin the night before a Swarthmore commencement speech. In contrast to Flip's preferred mode of transportation, Fredrick said van de Kamp often opted for convertible cars, including a red 1955 Pontiac.

The flourishes of both Fredrick and Lippincott showed admiration for their teacher and mentor. Van de Kamp died right about many things, wrong about two planet-sized things, and generally held in great esteem by the astronomical community.

Had he lived another five months, he would have lived to see the announcement that officially kicked off the era of planetary discovery: a real, verifiable planet around a Sunlike star.

The Champ

Many groups—including ones led by Marcy and Gatewood—were on the hunt for the first planet around a Sunlike star. Some candidates came and went. Others required dozens or hundreds of observations to officially confirm.

But an observation in January 1995 proved to be the real deal. Didier Queloz, a grad student at the University of Geneva, was working with his advisor, Michel Mayor, on the search for extrasolar planets via radial velocity, in other words, wobbles—just not astrometric ones.

Reportedly, his find was a chance coincidence. Out of a catalog of radial velocity signatures, he chose an F-type star called 51 Pegasi, roughly 50 light years distant. He was trying to calibrate his planet-finding code, opting for the star as one of a few promising candidates. It fell into place that night, a strong signal roughly every four days.

The planet completed one orbit around its star in four days, and radial velocity measurements placed its minimum mass near Jupiter—meaning the object was without a doubt a planet. While astronomers considered it *possible* to have such periods, it wasn't necessarily expected to find one in such a short period. "At this time, I was the only one in the world who knew I had found a planet," Queloz told the BBC in 2016. "I was really scared, I can tell you."[12]

There was some reason to be scared: if the past pages haven't nailed it in hard enough, finding a planet was then—and in some ways is still—really hard, and there were plenty of mistakes, ghosts, inexplicable data points, and other hiccups that never seemed to form a planet or a brown dwarf. Yet according to Queloz's data, the half-Jupiter-mass, quickly moving, ultra-hot planet was there.

Author Marcus Woo's article for *BBC Earth* also highlights one of the reasons radial velocity had been so difficult up until the 1980s. It was only then that a wobble going less than 1,000 kilometers per hour could be detected, while a planet like Jupiter might move the Sun 35 kilometers per hour, taking it from a rocket launch to residential driving. By the early 1980s, 54 kilometers per hour was finally possible to detect—which would have been enough to find 51 Pegasi b nearly fifteen years earlier than its actual discovery.

Much of the rest of 1995 was spent by Queloz convincing Mayor that he had truly found a signal, not an instrument error or other quirk of observing. Their paper was finally published in October 1995, five months after van de Kamp died.[13] Marcy followed up the observations and found that the Geneva team was on to something—he and collaborator Paul Butler were able to find the same signature at an entirely different observatory. This is one test that van de Kamp never passed. Almost all evidence for the Barnard's Star planets came from within Sproul, and the results were generally accepted—until other observatories actually took a closer look.

Depending on when you consider the official beginning of exoplanet searches to have kicked off—70 Ophiuchi observations in 1855 or See's observations in 1895, or perhaps (more logically) Kaj Strand's work on 61 Cygni in 1942—Queloz's discovery was either the culmination of 140, 100, or 53 years of efforts. Van de Kamp, who worked so hard at finding planets around main sequence stars, wasn't around to see the first confirmed planet around such a star.

Yet, 51 Pegasi would have been far beyond Sproul's observing program, even at a cosmic hop, skip, and jump distance. Smaller than any planet van de Kamp proposed, it would have been far outside the margin of error of the 24-inch refractor, which could only find planets, not stars. But that heralded an entirely different problem: the early days of exoplanet hunts meant astronomers finally had to figure out what was a planet and what was a failed star, because they were finally discovering both.

Rafael Rebolo of the Institute of Astrophysics in the Canary Islands and colleagues were looking for low-mass objects in the Pleiades star cluster and stumbled on an intriguing object in January 1994. Rebolo had been working on finding a brown dwarf throughout the 1980s, and identified a promising candidate in March 1990 with the object UX Tau C. But that object was young and right at the mass boundary between star and brown dwarf.

But this object first gave Rebolo the idea to look for lithium signatures in brown dwarf candidates. Lithium doesn't last long in the cauldron of a main sequence star, but it can be spread throughout a brown dwarf, as it wouldn't be destroyed by the heat of the star. UX Tau C had lots of lithium, but it was also very young—young stars may hold on to their lithium for a limited time before it is broken down into lighter elements. The next few years would be spent trying to identify low-mass objects and determining their lithium content. A few interesting quirks turned up, but nothing solid enough to pursue as a strong brown dwarf candidate.

But in January 1994, Teide 1 came along. Rebolo and his colleagues scrutinized the object, which sat in the Pleiades cluster, a grouping of several young stars moving together. It was still glowing red, so it might be a red dwarf . . . or it could be a young brown dwarf. Initial spectral measurements in October 1994 pointed toward a red dwarf, but December measurements were much more precise. They took the spectrum obtained and compared it to LHS 2065, a cool star in the same field. They were able to identify it as a very cool object indeed. But they needed radial velocity measurements and a better constraint on lithium burning. Because the Pleiades move together, mass measurements are possible. They also looked at back catalogs of the Pleiades in infrared wavelengths and found images of it going back to 1986. By compiling spectroscopic measurements, photometry, and astrometry, they were finally able to draw a conclusion: it was a solid brown dwarf candidate. They drafted a manuscript in the new year, submitting it to *Nature* on May 22, 1995.

Then, according to Rebolo, they became aware of another object. Researchers had discovered the very cool object PPl 15 in 1994, and it was jammed right between the mass of a brown dwarf and the minimum mass of a star, and there was *some* lithium in the atmosphere, which might happen in very young stars. But PPl 15 also seemed to be brighter than Teide 1. Researchers working on PPl 15 invited Rebolo and his collaborators to work together as their respective manuscripts awaited referee approval. Rebolo got word on August 14, 1995, that the paper had been accepted, and it was published on September 14, 1995. In November 1995, they went to Keck Observatory to get better spectroscopy readings, and they were able to confirm lithium in the atmosphere. It officially passed the litmus test of being a brown dwarf, the first one confirmed. This was the same month Mayor and Queloz announced their planet. In fact, the confirmation of lithium in the atmosphere of Teide 1 came on the exact same day (November 23, 1995).[14]

This came coincident with the Cool Stars IX meeting, where Ben Oppenheimer was presenting his results on Gliese 229B, *another* brown dwarf. As mentioned in the last chapter, it was discovered in 1988, but no one was quite sure what to make of it at the time because its mass put it at the star/brown dwarf boundary. The first three brown dwarfs had come in rapid succession, all around the same time as the first planet. Teide 1 gets the honorific of *first,* as it was the first to be verified as a brown dwarf, crossing over the candidate threshold. All three early brown dwarfs were nearer the star/brown dwarf boundary than the brown dwarf/planet boundary. A few other objects discovered in the 1980s and early 1990s would eventually be confirmed as brown dwarfs as well.

There used to be a sort of consensus: brown dwarfs formed like stars, from a gas cloud collapsing in on itself, and planets formed from dust leftover from this planetary formation. But the presence of planets that seem to form on their own in interstellar space—rogue planets formed like stars—throws one kink in that idea. Planets, by and large, form after a star has ignited or a brown dwarf has failed to ignite. (A small handful of brown dwarfs are known to have orbiting planets.) Dust in debris disks accumulates into larger and larger objects until a few of those can be considered planets. Some astronomers even liken the planetary formation process to a pebble-by-pebble accretion. But some low-mass brown dwarfs are in quite planet-like configurations, orbiting a star that's either its parent or its binary, depending on how you choose to view it. Brown dwarfs may not be the "missing mass" Jill Tarter was looking for when she named them that in the 1970s, but they are certainly a missing link between the two objects, sometimes exhibiting behaviors of both.

They're also really hard to detect since many of them emit little or no light, especially if they've since cooled down from their formation. Indeed, for years, Barnard's Star and Wolf 359 were considered to be the second and third closest stellar systems to Earth, but in 2013, Luhman 16—a pair of brown dwarfs—were discovered 6.5 light years away, driving a wedge between Barnard's Star and Wolf 359, and displacing Lalande 21185 to only the fifth closest system.

Luhman 16 A and B were found by the Wide-field Infrared Survey Explorer (WISE), which was primarily a solar system mission to find small objects in our backyard. But it has discovered several brown dwarfs, six of them within 20 light years of Earth, as well as 15 red dwarfs, some straddling the star/brown dwarf boundary.[15] This has added new, stranger worlds to our cosmic backyard—and hints that maybe more brown dwarfs are close to us, too dim even for WISE.

Between 1995 and 1999, 10 brown dwarfs were discovered. In that same span, 26 planets were discovered. Some were half Jupiter's mass, while others stretched up to eight times its heft. Of those, around 20 have an orbital period, or "year," less than that of Mercury. Twelve of them orbit in less than three weeks of Earth time. These close-in gas giants have come to be known as Hot Jupiters, owing to their proximity to their home stars.

The shortest period discovered in the 1990s was a three-day orbit for HD 1871123b. Since then, we've discovered planets with orbital periods of less than a day, the most extreme of which is PSR J12719-1438b, which completes an orbit every two hours around its home pulsar. It may not even be a planet in the classical sense—if any pulsar planet can be considered a "classical" sense planet. It's a hair over the mass of Jupiter, but its density compresses it into a more Uranus-like radius. The most likely scenario is that the pulsar's progenitor star was locked in an orbit with a Sunlike star. The cataclysmic explosion ripped off the outer layers of the Sunlike star, leaving behind a white dwarf that was further deprecated and eventually condensed into a giant hunk of carbon, meaning it could be a *diamond* planet. This may also make it the most unusual planet in the galaxy by most standards. When pulsars aren't forming zombie planets, they may be stripping stars of their stellar dignity and cutting them down into planet-like parameters. This further makes the line between a failed star and a planet murky.

Back to La-La-Lande

George Gatewood entered the planetary era with old school planet-hunting knowledge in a new world of verified discoveries. His astrometric work continued, mostly on nearby stars. Though MAP had limited abilities in finding planets—the scenario that plays out again and again in astrometric work—it was fantastic in refining the kind of work done at Sproul. It was one of the last ground-based astrometric programs devoted to hunting planets, though Gaia data are already yielding data on known exoplanets . . . and are hiding data on unknown ones, waiting for us to find our astrometric planets.

So it came with some surprise when Gatewood announced in 1996 that he had found not just a planet with MAP, but a whole slew of them. And they were around a familiar place: Lalande 21185. As you may recall, that star was the second planetary "discovery" at Sproul, one that fell out of favor once the true limitations of the refractor were understood. Ross 614B may have been Lippincott's career crown jewel, but, had the estimations been correct, Lalande 21185 wouldn't have been far behind.

Gatewood's announcement came just a half-year after the discovery of 51 Pegasi b. From MAP data, Gatewood claimed he saw what was unmistakably a wobble in the motion of Lalande 21185. His announcement—given at the fall 1996 AAS meeting—indicated that there were planets, plural, in the system.[16]

Gatewood was somewhat cautious about the claims, which is understandable given the history of the system. Lippincott's initial studies of the star indicated something roughly 10 times the mass of Jupiter orbiting the star, in contrast to Gustav Land's early work showing no deviation in the star's path around the galactic center. But Gatewood had disproved Lippincott's planet in 1973, shortly after he had evaporated van de Kamp's Barnard's Star claim. Russian radial velocity studies of the star published in 1992 found, at the very least, nothing like Lippincott's claim as well.[17] A 1992 study by Gatewood had also reported no planet-like wobbles in the motion of the star.[18] (Oddly, a 1983 paper using infrared interferometers claimed to have seen a planet at Lalande 21185 and a handful of other stars—though they all ended up being either nothing or previously unseen binary stars.)[19]

The May 1996 announcement of at least one planet around the star was a little surprising—just four years earlier, Gatewood had ruled out planets there, and now he was announcing up to *three* planets there, which is quite an about face. The first proposed planet—and the one with the most solid backing—was a gas giant about 90 percent the mass of Jupiter, locked in a 5.8-year orbit, and a second, more ambiguous object, in a 30-year orbit, plus initial indications for a third object of even more ambiguous provenance.

Gatewood acknowledged the ambiguity of the observation, telling *Pitt Magazine*, "I know there's something orbiting this star. I know it with a great deal of certainty. The details I don't know as well. Two planets are possible answers." In the same feature, Anita Cochran of the University of Texas at Austin told reporter Tommy Ehrbar, "It's not very definitive, but Gatewood's as good as anybody in the field. I would just like to see more data." Another astronomer, David Black, even remarked that Gatewood's announcement was "more like what a planetary system should be, if that's what it is."[20]

But by the end of 1997, Gatewood had withdrawn the discovery, saying he'd found an "acceleration" of the star rather than an orbit. Paul Butler reportedly had a hand in correcting Gatewood's estimations of the "planets."[21] But Gatewood also told Bruce Dorminey in *Distant Wanderers* that he was merely retracting the 30-year-period observation but believed there was a 6-year-period planet there—more similar in form to Lippincott's 8-year orbit (which, confusingly, Gatewood had punted to the 30-year orbit.)[22] But in the end, none of Gatewood's discoveries at Lalande 21185 have been confirmed.

Interestingly, the Russian researcher who failed to turn up a companion at Lalande 21185—Natalia Shakht—reported a possible 4 Jupiter-mass planet around 61 Cygni in 1998, though it wasn't the first time the Russians had made such proclamations around the stars. She also suggested substellar objects around ADS 11632 (which had been suspected to have a low-mass star orbiting it previously) and reported an object around ADS 12815—which does have at least one planet of vastly different parameters than those reported from Shakht's Pulkovo Observatory.[23] Shakht has continued to push for a 6.5-year large planet or small brown dwarf in 61 Cygni. Were this true, it might vindicate Strand's work, as he proposed an 8-ish Jupiter-mass object in a 5-year orbit with less precise instruments. But few, if any, Western astronomers have corroborated the Pulkovo work.

The most recent paper from Shakht regarding 61 Cygni indicates that future programs like the Hunt for Observable Signatures of Terrestrial [planetary] Systems (HOSTS) could find a planet there. Indeed, 61 Cygni is on the short list of target objects of that program, which will use the Large Binocular Telescope Interferometer to look for planets around nearby Sun-type stars.[24] The HOSTS system will look for gaps in the glow of debris disks around a handful of stars to infer the presence of planets. Vega is also on the list—which would be an exciting find for astronomers—as is Fomalhaut.

In 2017, Paul Butler was ready for an announcement, though. It wasn't a done deal, but he'd reportedly discovered a planet around Lalande 21185—a hot, hot world in a near-10-day orbit. It also wasn't a gas giant—it was a planet known as a super-Earth, a class of large rocky or small gaseous planet a few times Earth's mass. Butler estimated a 3.5 Earth-mass planet. Such worlds were far beyond the technology of 1996 and were definitely astrometrically out of the question from the ground, so it's no surprise that Gatewood hadn't detected it. (It should also be emphasized that Butler only considers it a *candidate,* rather than a confirmed planet.)[25]

Butler's discovery was greeted with little fanfare—in April 2017, versus May 1997, we knew of thousands of *confirmed* planets and plenty of hot super-Earths like the putative planet Lalande 21185b. A candidate planet is nothing special right now—even though the candidate is *nearby,* it's not habitable, and Lalande 21185's history is more of a trivia question than an astronomical white whale. We have nearly 5,000 candidates from the Kepler spacecraft plus hundreds more from other programs. Butler's discovery involved 20 years of radial velocity observations. At the time of this writing, the planetary candidate discovery is less than a year old, so it may not be long before it's confirmed. For instance, the Arecibo Observatory in Puerto Rico is analyzing data (taken prior to the hurricane that damaged much of the island) that include several observations of Lalande 21185. Barnard's Star is also on its target list.[26]

Gatewood's discovery ran through the news cycle, but in the end, it never moved past a conference paper. Perhaps most of the evidence had evaporated by the time he would have published his findings, or David Black coached him out of it. And perhaps Gatewood, based on years of experience, knew better than to belabor the eventual fallout.

It wasn't the last time he found an object in old Sproul targets, though. Gatewood was part of a group—which also included one of the scientists who worked on Gamma Cephei, Stephenson Yang—that detected a planet around Epsilon Eridani.[27] The planet is far, far from the one envisioned by van de Kamp as a star-planet. It has around the mass of Jupiter, rather than six Jupiters, and it didn't orbit in 25 years. Its orbital period is instead probably around 7 years long—plus or minus a few Earth months. For the most part, this planet, which was found via Hubble data, is regarded as a strong candidate, though the presence of a debris disk means it's not a 100 percent bona fide planet.

Gatewood also worked behind the scenes in the planetary community, assisting other universities in getting the resources they needed to look for extrasolar planets, and worked on attaining mass measurements for already established planets.

Redemption

In the early 2000s, 51 Pegasi b was unseated. Sort of. While a *Sky and Telescope* article from the 1990s suggested that 51 Pegasi b may have been a quirk of the star's atmosphere, this was something a little bit different.[28] Indeed, time after time, 51 Pegasi has stood up to scrutiny as the first verified planet. But as it ends up, Gamma Cephei Ab was the first planet discovered after all.

At the time of Campbell, Walker, and Yang's 1988 study, radial velocity was finally emerging as a planet-finding method but was still just primitive enough to be prone to errors. The problem of the original detection came not from an instrument error per se, but because Gamma Cephei A's mass was poorly understood, which also led to some mismeasurements of Gamma Cephei B. After new masses were determined, better software than that available in 1988 helped further separate signal from noise. There was, indeed, a movement every 906 days, a periodic bump in the radial velocity data caused by something twice Jupiter's mass.

Sadly, in the intervening time, the fallout of the Gamma Cephei affair had driven Campbell out of academia. His work simply wasn't generating the kind of grant funding he needed to stay on with the program, and even a cash injection from Carl Sagan couldn't manage to keep him at UBC—Campbell was still technically a postdoc, and UBC regulations required the funding to go to tenured faculty. Dejected and frustrated, Campbell deleted nearly all his work while at UBC. Of course, at the time, it seemed fruitless, even foolhardy. He left astronomy to write tax software, something more steady and able to provide for his growing family.

The data deletion set back the UBC staff a year, during which Walker and Yang had to painstakingly reconstruct what they could of the data. This nearly caused Walker to reappraise Gamma Cephei in

1992 before ultimately concluding his evidence for a planet was too weak. The paper that was ultimately published—which nearly pushed for a planetary hypothesis—was printed in June of that year.[29] In 2002, Gamma Cephei Ab was restored to its full planetary status by new data. Had the researchers in 1988 had better software and mass measurements, it would have easily qualified as the first discovered exoplanet.[30]

It took another decade to have enough data on HD 114762b to understand its true mass. In 2012, new data confirmed the mass at around 11 Jupiter masses. This is generally in line with a planet—a really heavy one, but a planet nonetheless.[31] (Given the blurring view between giant planet and small brown dwarf, though, this could be reappraised in coming decades if we're ever able to know more about its atmosphere and composition.)

The first planets were discovered in 1988 and 1989—we just had no idea what we were looking at back then.

One at a Time vs. All at Once

The early days of exoplanet detections were marked by their focus on star-by-star analyses, almost always through radial velocity, making radial velocity one of the most successful methods for finding exoplanets, discovering 746 worlds as of March 2018. Compare that to 90 found through some form of direct imaging (which is limited to large, hot, and young planets) or 67 through microlensing—chance happenstance when a heavy object passes in front of a background star and acts like a giant magnifying glass. Those are the third and fourth most successful methods for finding exoplanets.

But far and away the most successful approach has been the transit method of finding exoplanets. It's found 2,789 in all within the same timeframe as the radial velocity headcount. There are 3,705 planets out there, so transiting planets make up 75 percent of all planets discovered. But around 2,648 of those 2,789 planets have been found by one spacecraft: Kepler. NASA's mission stared at one patch of sky for four years and found 2,341 of those. It subsequently found another 307 after the spacecraft's stabilizing reaction wheels went out of commission. The craft was nearly written off before astronomers figured out they could use sunlight to balance the craft and focus on finding large worlds.

Around 30 of the 2,341 Kepler planets are potentially habitable—places where we could look for water with the right instrumentation. Answers to whether or not red dwarf planets can retain oceans will soon finally be answered. More than 5,000 candidate planets await confirmation from the craft, which will require intense, star-by-star study or a better data reduction method.

If you take out the worlds discovered by Kepler, we have only 1,000 planets to work with. That's because Kepler served as a survey of just one small patch of sky, counting as many planet transits as it could. Previous surveys dealt with a few dozen stars at a time—if that. Kepler, if nothing else, showed that planets weren't rare at all, and there are millions—or trillions—out there, waiting for our discovery.

At this point, aside from a few high-profile flubs, it would be hard to bean count the number of planets that have evaporated under further scrutiny. One early retraction of note is Rho Coronae Borealis, announced as an exoplanet found by radial velocity in 1997. But astrometric measurements made it clear that the radial velocity signature had vastly underestimated the object's mass. It wasn't a 1.1 Jupiter-mass planet—it was a star of more than 100 Jupiter masses. Hipparcos data revealed the starlike nature of the companion. It was, for our purposes, also one of the only planets *disproved* by astrometry.

There's also the incredibly contentious Gliese 581 system. Announced in 2007, it held five planets, including a few in the habitable zone of its home star. But the most habitable of the worlds—Gliese 581d—may not exist at all. It has rarely held up to further scrutiny.

And, of course, in 2012, a short-period exoplanet was announced in the Alpha Centauri system but was quickly disproved by the astronomical community. Another team spent the next few years concentrating on Proxima Centauri, and after laborious work, under a thick veil of secrecy, they were able to confirm at least one planet that is the closest exoplanet system to Earth—and will be for the next 36,000 years, when Ross 248 moves in closer.

That team—calling itself Pale Red Dot—later rechristened themselves into Red Dots. Their work has switched focus to continued Proxima Centauri observations but added in Barnard's Star and Ross 154. And in November 2018, they found something tantalizing at Barnard's Star—but more on that in the next chapter.

In a 2018 conference talk, MIT's resident exoplanet expert, Sara Seager, mentioned that exoplanet astronomy is, in some ways, taking a turn back toward its beginnings. There will still be some large-scale surveys, but those will be intended to find a handful of candidates for future studies. Other projects like Red Dots will focus on a few stars at a time. This is partly because, with much of the heavy lifting done on censuses of stars, we are at the edge of being able to know previously unfathomable details about planets—and we may be studying them one by one with giant telescopes and better optic technology. Barnard's Star will be a prime target . . . though as we'll get to in the next chapter, there's finally a glimmer of hope.

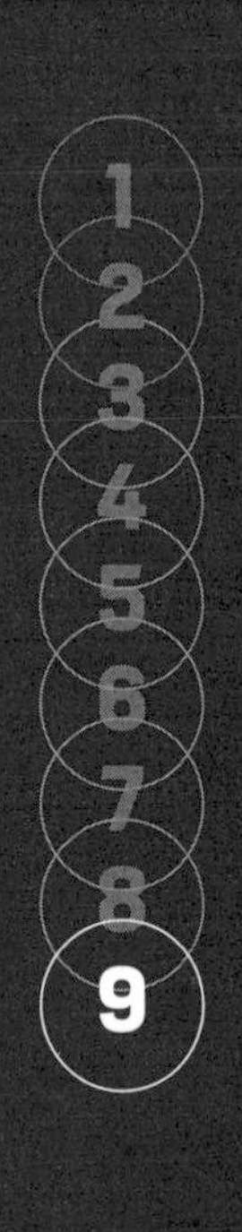
1
2
3
4
5
6
7
8
9

On the grounds of Swarthmore, a fence surrounds what was once the Sproul Observatory. It's a wet October day, and the construction crew is at lunch. The 24-inch refractor has been gone for a few months after more than 100 years at Swarthmore.

The group NWA Space (short for Northwest Arkansas) bought the scope and are planning on installing it at their own observatory, with a little help from Supporting STEM and Space. The refractor will mostly be used for K–12 education.

The observatory hadn't seen the light of night in quite some time. The building is being refurbished into a visitor's center, leaving no room for the old instrument on Swarthmore soil.

But the days of astronomy are far from numbered. In 1998, Eric Jensen joined the Astronomy Department and used the refractor—in its final years—to look for planet-forming regions. That's the same year Wulff-Dieter Heintz retired, though, like van de Kamp, he hung around for a few years after to continue his work.

Heintz's retirement was, perhaps, the official end of the van de Kamp era. No one from the Barnard's Star affair was left at the observatory. Lippincott still comes around, Jensen tells me. No one I have spoken to quite knows where John Hershey is—not his colleagues at the Space Telescope Science Institute, nor anyone at Swarthmore. His last publication came in 2006, and he's stayed largely out of the public eye since then—if he's alive.

As for the Barnard's Star affair? Jensen says, "It's acknowledged as part of the college's history. It's past the time where it was personal for anyone here anymore."

There are a few students who have gone on to huge careers in astronomy. Sandra Faber was awarded the National Medal of Science in 2013. Christopher Chyba, a 1982 graduate, went on to become a notable astrobiologist. John Mather, a 1968 physics graduate, was listed as one of the 100 most influential people in the world in 2007, thanks largely to his work at NASA Goddard. He's also one of the senior scientists on the James Webb Space Telescope, the tennis-court-sized follow-up to Hubble. Hubble would not have been possible without Nancy Grace Roman, another Swarthmore grad from early in van de Kamp's career. Bruce Draine became well known for his work on cosmic dust and was inducted into the National Academies of the Sciences in 2007.

Endings

On Halloween 2000, Kaj Strand passed away at the age of 93. He was, at the time, living at the Manor Care Nursing Home in Bethesda, Maryland. By then, he had been retired from the Naval Observatory for 23 years. Unlike van de Kamp, he'd lived to see the exoplanet era finally officially emerge. Obituaries of the time made scarce mention of 61 Cygni. Perhaps it was because his work moved far beyond looking for planets, into more orthodox work on double stars.

Heintz died in 2006 of lung cancer, leaving behind an intense legacy on double and binary stars. Heintz had been a lifelong smoker and had survived rough and tumble beginnings in Hitler's Germany. He was 76 at the time of his death—significantly younger than Strand or van de Kamp had been at the time of their deaths.

In his last years at Swarthmore, Heintz built out his work on binary and double stars substantially—perhaps hoping that it would serve as his lasting legacy. He also coauthored a few papers on variable stars, those whose brightness vastly changes over a short time scale.

As of the date of this writing, Sarah Lee Lippincott is still alive—spry even, Jensen says—but efforts to contact her failed. Attempts to reach John Hershey failed as well. As noted above, he stopped publishing after 2006, when he seemingly left the Space Telescope Science Institute. His colleagues from the time—surprisingly few—are unaware of his whereabouts.

George Gatewood retired from the University of Pittsburgh in 2009. His last paper dealt with the parallax measurements of several nearby stars. Gatewood—perhaps enjoying his retirement—did not return an interview request.

New Beginnings

Around 2004, the Sproul Observatory closed its doors. The astronomy faculty moved into the Science Center on the Swarthmore campus, awaiting the installation of a new observatory. It was finished in 2008, once funding had been secured.

"If you were building a new observatory from scratch and looking for the best site for it, you probably wouldn't choose the suburbs of Philadelphia," Jensen told me on one of my many visits to the Swarthmore campus.

That's perhaps why the focus has shifted, just a little.

The new observatory bears a familiar name. Once it was finished, it was christened the Peter van de Kamp Observatory. Van de Kamp's awards are hanging on its walls, and a framed painting of van de Kamp glowers at undergrads. One graduate of Swarthmore I spoke to—my colleague Josh Sokol—remembers the visage watching over him on observing nights. The serious expression in the painting is perhaps a little in contrast to the goofy Chaplin fan with a little bit of a flamboyant side, who was also a deadly serious astronomy scholar.

"It's nice to have his name on the observatory," Jensen says. "Although he's best known for the Barnard's Star stuff and that didn't work so well, he did a lot of really fundamental science. He really anchored our understanding of low-mass stars."

A lot has changed between the end of Sproul and the beginning of the Van de Kamp Observatory. Swarthmore's astronomy program became less focused on astrometry but didn't shift its focus away from the potential of finding planets. But now, it does it through photometry.

The Van de Kamp Observatory is one of several collaborators on the Kilodegree Extremely Little Telescope, or KELT, program. Two finder observatories, one in Arizona and one in South Africa, stare at the dark, dark night sky in their locales. KELTs are not much more than a high-end consumer camera fixed on a telescope mount. Those two facilities look for evidence of transiting planets. Then the other KELT collaborators step in. This time, Swarthmore isn't mostly alone in the planet search. KELT is spearheaded by Ohio State University, Vanderbilt, Lehigh University, Penn State, the Harvard–Smithsonian Center for Astrophysics, and a few other observatories that follow up on the small blinks captured by the finder scopes.

Along with KELT, the Van de Kamp Observatory is also part of the Young Exoplanet Transit Initiative (YETI) and plans on performing follow-ups to the Transiting Exoplanet Survey Satellite, or TESS, NASA and MIT's follow-on to Kepler. Unlike Kepler, TESS will focus on bright nearby stars.

Those, as it ends up, are the kinds of stars that KELT—and the Van de Kamp Observatory—are best attuned to. On a casual night, the Van de Kamp Observatory can see maybe to the 12th magnitude. That's barely enough to see Barnard's Star. Such has the sky changed in Philadelphia.

So now, the observatory stares at bright stars. Very bright stars. Some of the brightest in the galaxy. As Jensen says, the planetary hunt started with stars like the Sun but gradually moved in one of two directions: you either went for brighter and hotter, or smaller, dimmer, and cooler, like Barnard's Star.

And here's the thing: Swarthmore has found planets now. KELT has discovered 18 planets and a brown dwarf. Swarthmore observed 8 of those worlds. It started out in a similar way as the early exoplanet hunt: their first candidate, KELT-1b, ended up being a brown dwarf. It has a 29-hour orbit, and like many of KELT's worlds, is extremely hot.

KELT-9b really brings the heat. It's the only planet discovered so far around a B-type star, the second most massive type of star in the universe. KELT-9 is about 650 light years away and burns in hot blue. KELT-9b orbits in just a day and a half, sitting merely 3,216,271 miles from its home star. One side of the planet faces its star at all times, which brings its outer gas layers to an astounding 7,770 degrees Fahrenheit. That's way warmer than a red dwarf and just a hair cooler than the Sun, and at first, the astronomers weren't sure if they were seeing a planet at all. Two researchers even placed a mutual bet on a nice bottle of scotch if it ended up being a planet.[1]

Swarthmore is thus finding planets—they're just at the opposite ends of the spectrum from those hunted by van de Kamp and company, using an entirely different method.

The average observing night is entirely different at Swarthmore now, too. Where an astronomer might have stayed up all night in the

1960s, tracking their telescope and exposing plates, now, the Van de Kamp Observatory staff and students simply set up an observing campaign via computer code. They work on following up the KELT candidate catalog, choosing to point the telescope at a star when they think a planet might transit, then moving on to the next candidate. Once the observing campaign is put in for the evening, then the astronomer can simply walk away, and the telescope does all the work. "That lets us actually sleep at night," Jensen says.

Perhaps it's a less romantic astronomy process than, say, the ones depicted in an illustration of Flip in a wizard's hat staring through the Sproul eyepiece, as one doodle sent to Lippincott displayed. There's no less of a fascination with stars within the department, of course. But astronomy today is largely driven by code and automation. With the right instrument, amazing amounts of automated science can be done.

It's not as if no one is staring up at the sky from Van de Kamp Observatory. You can still take out the CCD sensor from the telescope, put in an eyepiece, and put together an impressive public viewing night. Every second Tuesday of the month, the observatory opens up, and the public comes in. But ask astronomers today—not just at Swarthmore, but at almost any observatory in the nation—what gives them headaches, and many will answer that it's as much the Python code they input to run their observing scripts as it is any part of the actual telescope.

It's simply a different time from the olden days, if we dare call van de Kamp's time that.

Perhaps one thing isn't different between Sproul then and the Peter van de Kamp Observatory now: the focus, based on the limitations of the instrument, is on large, Jupiter-mass planets. The smallest KELT planet is still 60 times the mass of Earth and 5 times the mass of Neptune. The periods may be shorter, but the hunt for massive planets still drives the program.

The day I visited Jensen, a student was putting in the night's routine. I asked what the target was for that night. She rattled off a long catalog input number. There are innumerable objects in the sky, and so candidate catalogs for objects give them indistinct and hard-to-remember names. For instance, Kepler's catalog gave the input number KIC 8462852 to what became one of its most interesting and intensely scrutinized objects: a star dimmed by as much as 20 percent, far larger than a planet, and astronomers were at a loss as to why—and in the popular press, it was played up as a possible alien megastructure. But KIC 8462852 doesn't roll off the tongue, so most people took to calling it Boyajian's Star.

The catalog number for this KELT object is . . . long. So long I got lost trying to write it down. It's also an oddly behaving object. Sometimes whatever is transiting will show up 15 minutes earlier than it did last time, or 4 minutes later. The KELT team are still trying to figure out what they're looking at.

"This one has been observed a couple times," Jensen says. "It's a shallow signal on the edge of what we can detect."

Onward and Upward

Whereas in the 1950s through the 1960s, Sproul was the only observatory really looking for planets, and the 1970s brought out Allegheny but not a lot of other competition, KELT is one of dozens and dozens of ground-based exoplanet projects, all with varying degrees of goofy acronyms. HARPS, HAT, and WASP are the big ones that have discovered dozens or hundreds of exoplanets. OGLE, KELT, and the not-that-fun acronym Anglo-Australian Planet Search (AAPS) are middle tier. ESPRESSO, SPECULOOS, and SEEDS haven't found a lot of planets, but they may make you hungry. And those are just a few examples.

Perhaps the most important ground-based discovery in the last few years is the TRAPPIST-1 system.

TRAPPIST-1 is a small, cool star—not much larger, radius-wise, than Jupiter, making it far smaller, cooler, and dimmer than Barnard's Star. It's only 39 light years away, but at 18th magnitude, it would have been far outside the gaze of the Sproul telescope, going undiscovered until 1999. It wasn't until 2016 that astronomers realized there were planets orbiting the star. And it wasn't until 2017 that they realized there were seven of them, all roughly the mass of Earth, some a bit bigger, some a bit smaller. Under the right conditions, all seven could have some form of water, and at least three could be hospitable to life. There are initial suggestions that there's water in the system, though we'll learn much more when the James Webb Space Telescope launches.

The TRAPPIST telescope is two Sproul-sized aperture telescopes working together in Chile to find the dimmest stars in the universe. It has discovered something Peter van de Kamp never could have envisioned—seven planets swirling around their star in near-synchronous orbits, creating a spectacular system.

Around 24 planets are nearer than 40 light years from Earth. There's one around a star in the wrong hemisphere for van de Kamp to have ever studied, Proxima Centauri. It's the closest one. After that is Epsilon Eridani's planets, in a system well scrutinized by the Sproul program.

For the longest time, Barnard's Star wasn't on that list. But in late 2018, the Red Dots team announced, finally, a frigid super-Earth in the system. It's at the snow line of Barnard's Star, the zone where water, carbon dioxide, and other chemicals condense into solids. This places it far outside the habitable zone of the star. The planet, around three times the mass of Earth, crawls around Barnard's Star in 230 Earth days, making it something of a rare find in an era when many planets have "years" that last less than a week, due to statistical bias in how easy they are to find in a dataset. The researchers took 20 years' worth of data to find the planet via radial velocity, that subtle process of finding how much a planet displaces its star. Something like a super-Earth has a much weaker tug compared to close-in gas giants, the kinds initially discovered. Of course, a frigid world isn't quite what any astronomer hopes for when searching for life around nearby stars. There seems to be nothing Earth size or larger in between the super-Earth and Barnard's Star.

We could, some day, see this world. NASA has discussed using a *coronagraph*—an object that blots out the light of a star to look for planets to image directly—in upcoming space missions, like WFIRST, though its future is unclear as of this writing. But the distance from the star, the closeness to Earth, and the dimness of Barnard's Star makes this big ice ball a good target for imaging directly. So it's a planet we might actually *see.*

It's also a crown jewel in many ways, the confirmation that there is, at least, something at Barnard's Star. Up to now, it's never been a foregone conclusion. Several studies have ruled out anything like van de Kamp's planets at the star, precluding anything Jupiter to Saturn size.

But there's something quite mysterious in the dataset. The study's lead author, Ignasi Ribas of the Institut de Ciencies de l'Espai, was hesitant to quite make the comparison here, but there is the possibility of another planet, with a period of more than 6,000 days. If you don't have a calculator in front of you, that is greater than 15 years—an oddly van de Kamp–like period. "These variations are of similar period to the planets proposed by van de Kamp but their properties are not fully compatible," he told me. Their best guess is an object roughly the size of Neptune, spaced about 4 AU from Barnard's Star. (Jupiter is between 4.5 and 5 AU from the Sun.)[2]

Perhaps fittingly, the best way to confirm this weak radial velocity signal is an intense astrometric study, the kind that can't be done from the ground—if a Neptune-sized signal holds true, Gaia should be suited to the task. But Ignasi says the data from the mission aren't ready, as of fall 2018 to yield much in the way of planetary data. Those data releases will come closer to 2020.

It's unlikely, ultimately, that the astrometric signature will be very van de Kampian either. There's considerable doubt that he could see anything like this at all, whether at Barnard's Star or at 61 Cygni. That kind of gas giant planet may be startingly rare around these types of stars. A 2018 paper described the third transiting gas giant circling a red dwarf ever found. That planet, NGTS-1b, is 1/5 the radius of its own star and has sparked debate over how such planets are even theoretically possible.[3] Kepler-45b and WASP-80b are the two other known transiting gas giants, and only 7 percent of red dwarfs are expected to have a planet more massive than Jupiter in orbit around them. NGTS-1b is about 80 percent the mass of Jupiter; the other two are only half the mass of Jupiter. A few other gas giants around red dwarfs—such as Saturn-mass Gliese 649b, two Jupiter-ish mass planets around Gliese 849, and more-massive-than-Jupiter planets around Gliese 317 were found by other methods. A Jupiter-mass world is a tall order, let alone a star-planet. But we may find a Saturn-mass world or two in the mix somewhere.

The ongoing fascination with the Barnard's Star system speaks to the power that van de Kamp's planets have sparked in the popular imagination. When I spoke to astronomer Cullen Blake of the University of Pennsylvania in 2014 regarding his upcoming telescope, Minerva-Red, which is specifically designed to capture transits around red dwarfs, his first target was Barnard's Star. The telescope is mostly meant for red dwarf stars within 12 light years of us, close enough for glancing at Lalande 21185, Ross 248, 61 Cygni, Luyten's Star, or any of the numerous other stars on the Sproul shortlist of planets. Maybe, as Red Dots did, he'll find something at one of these places, something distinctly un-van de Kampian, but marking success in these systems. (Of course, Luyten's Star has two known planets, and Lalande may already have one.)

A 2014 study found that statistically, nearly all small stars should have a planet of *at least* the mass of Earth, and that one in four to one in five of them should have something between 3 and 10 times the mass of our home planet.[4]

Fifty-five years after van de Kamp first made his announcement, we at least know that Barnard's Star isn't alone out there. Maybe it has that distant Neptune making a slow orbit, that un-van de Kamp–like world in that van de Kamp–like period. Perhaps it has tiny Mars-like worlds evasive to today's technology, hiding closer in than the snow line. But it isn't alone, as we thought for so long.

There are planets there. Just not as we thought we knew them.

Notes

Introduction

1. Alberto A. Martínez, "Was Giordano Bruno Burned at the Stake for Believing in Exoplanets?" *Scientific American*, March 19, 2018, https://blogs.scientificamerican.com/observations/was-giordano-bruno-burned-at-the-stake-for-believing-in-exoplanets/.

Chapter One

1. Kaj Strand, "61 Cygni as a Triple System," *Publications of the Astronomical Society of the Pacific* 55, no. 322 (February 1943): 29–32.
2. Peter van de Kamp to William Leonard Laurence, January 21, 1943, Astronomy Department Records, 1899–1986, Swarthmore College Archives, Swarthmore, Pennsylvania (cited hereafter as Astronomy Department Records, SCA).
3. Dorrit Hoffleit, "Big Planet or Little Star?" *Sky and Telescope* 2, no. 3 (January 1943): 15.
4. James Hickey, "Denies Naming Planet Osiris," *New York Sun*, January 21, 1943.
5. James Hickey to Peter van de Kamp, January 21, 1943, Astronomy Department Records, SCA.
6. Dirk Reuyl and Erik Holmberg, "On the Existence of a Third Component in the System 70 Ophiuchi," *Astrophysical Journal* 97 (January 1943): 41.
7. Kaj Strand to Peter van de Kamp, April 19, 1943, Astronomy Department Records, SCA.
8. Kaj Strand to Peter van de Kamp, July 4, 1943, Astronomy Department Records, SCA.
9. Peter van de Kamp to Kaj Strand, July 14, 1943, Astronomy Department Records, SCA.
10. Jan Schlit to Dirk Brouwer, October 10, 1947, Astronomy Department Records, SCA.
11. Henry Norris Russell, "The Orbit of 70 Ophiuchi," *Publications of the Astronomical Society of the Pacific* 55, no. 323 (April 1943): 104.
12. William Stephen Jacob, "On Certain Anomalies Presented by the Binary Star 70 Ophiuchi," *Monthly Notices of the Royal Astronomical Society* 15–16 (1855): 228.

13. Thomas Jefferson Jackson See, "Researches on the Orbit of 70 Ophiuchi, and on a Periodic Perturbation in the Motion of the System Arising from the Action of an Unseen Body," *Astronomical Journal* 16, no. 363 (1896): 17–23.

14. Thomas Jefferson Jackson See, "Remarks on Mr. Moulton's Paper in A.J. 461," *Astronomical Journal* 20, no. 464 (June 1899): 56–59.

15. C. J. Peterson, "Thomas Jefferson Jackson See and the University of Missouri," *Bulletin of the American Astronomical Society* 14 (March 1982): 622.

16. Thomas Shirrell, "A Career of Controversy: The Anomaly of T. J. J. See," *Journal for the History of Astronomy*, February 1999, 25.

17. Kaj Strand, "The Orbital Motion of 70 Ophiuchi," *Publications of the American Astronomical Society* 10 (1946): 70.

18. Peter van de Kamp interview by David DeVorkin, March 18, 1979, Oral Histories, Niels Bohr Library and Archives, American Institute of Physics, https://www.aip.org/history-programs/niels-bohr-library/oral-histories/4929-2.

19. Van de Kamp interview, March 18, 1979.

20. Kaj Strand interview by David DeVorkin and Steven Dick, December 8, 1983, Oral Histories, Niels Bohr Library and Archives, American Institute of Physics, https://www.aip.org/history-programs/niels-bohr-library/oral-histories/23026-1.

21. Eric E. Mamajek, Scott A. Barenfeld, Valentin D. Ivanov, Alexei Y. Kniazev, Petri Väisänen, Yuri Beletsky, and Henri M. J. Boffin, "The Closest Known Flyby of a Star to the Solar System," *Astrophysical Journal Letters* 800, no. 1 (February 12, 2015): L17.

22. Peter van de Kamp to Maud Makemson, November 13, 1950, Astronomy Department Records, SCA.

23. Gerard Kuiper to Peter van de Kamp, January 7, 1940, Astronomy Department Records, SCA.

24. Peter van de Kamp to Gerard Kuiper, May 20, 1938, Astronomy Department Records, SCA.

25. Peter van de Kamp, "Stars or Planets?" *Sky and Telescope* 4, no. 38 (December 1944): 5.

26. Sarah Lee Lippincott to Peter van de Kamp, October 2, 1942, Astronomy Department Records, SCA.

27. Sarah Lee Lippincott and Peter van de Kamp, "A Determination of the Parallax and Mass-Ratio of δ Equulei," *Astronomical Journal* 51 (September 1945): 162.

Chapter Two

1. Peter van de Kamp to Kaj Strand, June 2, 1953, Astronomy Department Records, SCA.

2. Sergio Dieterich, Todd J. Henry, Wei-Chun Jao,Jennifer G. Winters, Altonio D. Hosey, Adric R. Riedel, and John P. Subasavage, "The Solar Neighborhood XXXII: The Hydrogen Burning Limit," *Astronomical Journal* 147, no. 5 (March 24, 2014), https://doi.org/10.1088/0004-6256/147/5/94; and Alexander von Boetticher, Amaury H.M.J. Triaud, Didier Queloz, Sam Gill, Monika Lendl, Laetitia Delrez, David R. Anderson, et al., "The EBLM project III: A Saturn-Size Low-Mass Star at the Hydrogen-Burning Limit," *Astronomy and Astrophysics* 604 (2017), arXiv.org e-Print archive, July 12, 2017.

3. Ann Ewing, "Lightest Weight Star Found by Biggest Telescope," *Science News Services*, April 4, 1955.

4. Willem Luyten, "On the Orbit and Mass of L726-8," *Publications of the Astronomical Society of the Pacific* 66, no. 393 (December 1954): 337–338.

5. Meghan Bartels, "How Harvard's Vast Collection of Glass Plates Still Shapes Astronomy," *Astronomy*, February 1, 2017, http://www.astronomy.com/news/2017/02/stars-frozen-in-time; see also Dava Sobel, *The Glass Universe* (New York: Viking, 2016).

6. Sarah Lee Lippincott to Leslie Sevy, March 19, 1965, Astronomy Department Records, SCA.

7. Adolph Katz, "Woman Astronomer at Swarthmore a Champion of Discovering Stars," *Philadelphia Bulletin*, March 16, 1960, 25.

8. "Report of the Visiting Committee to the Swarthmore College Astronomy Department," March 1975, Astronomy Department Records, SCA.

9. Peter van de Kamp to Kaj Strand, February 5, 1968, Astronomy Department Records, SCA.

10. Peter van de Kamp to Dr. Courtney C. Smith, president, Swarthmore College, June 22, 1961, Astronomy Department Records, SCA.

11. "Sarah Lee Lippincott: She's a Star Studier," *Science World*, April 18, 1974.

12. Sarah Lee Lippincott to Mildred Shapley Matthews, July 5, 1962, Astronomy Department Records, SCA.

13. Sandra Faber interview by Alan Lightman, October 15, 1988, Oral Histories, Niels Bohr Library and Archives, American Institute of Physics, http://www.aip.org/history-programs/niels-bohr-library/oral-histories/33932.

14. Richard Hollingham, "How a Nazi Rocket Could Have Put a Briton in Space," *BBC Future*, August 25, 2015, http://www.bbc.com/future/story/20150824-how-a-nazi-rocket-could-have-put-a-briton-in-space.

15. Joseph Myler, "A Mere Trillion-Miles Up There," *Harrisburg (PA) Evening News*, November 18, 1955.

16. United Press, "Astronomers Think Other Worlds Might Be Discovered," *Buffalo (NY) Courier Express*, October 2, 1955.

17. S. S. Kumar, "On the Nature of the Planetary Companions of Stars," *Zeitschrift für Astrophysik* 58 (1964): 248.

18. S. S. Kumar, "Planetary Companions as Late-Type Stars," in *Proceedings of the Colloquium on Late-Type Stars*, ed. Margherita Hack (Trieste: Osservatorio Astronomico, 1967), 341.

19. Benjamin Adelman, "Do Stars Have Planets?" *Junior Astronomer* 5, no. 2 (October 1955): 2–4.

20. Peter van de Kamp and Sarah Lee Lippincott, "Astrometric Study of Lalande 21185," *Astronomical Journal* 56 (April 1951): 49.

21. Sarah Lee Lippincott, "The Unseen Companion of the Fourth Nearest Star, Lalande 21185," *Astronomical Journal* 65 (August 1960): 349.

22. Sarah Lee Lippincott, "Astrometric Analysis of Lalande 21185," *Astronomical Journal* 65 (September 1960): 445–448.

23. Science Service, "'Nearby' Object May Be a Planet," *Austin Statesman*, May 5, 1960.

24. *Science and Mechanics*, October 1960, 76.

25. Stuart Brown, "Scientists Search for Signs of Life in Outer Space," *Philadelphia Bulletin*, October 9, 1960.

26. Associated Press, "Old Sol May Have Twin, Astronomer Maintains," *Tucson Star*, June 14, 1961.

27. Sarah I. Sadavoy and Steven W. Stahler, "Embedded Binaries and Their Dense Cores," *Monthly Notices of the Royal Astronomical Society* 469, no. 4 (August 21, 2017): 3881–3900.

Chapter Three

1. Peter van de Kamp, "Astrometric Study of Barnard's Star from Plates Taken with the 24-Inch Sproul Refractor," *Astronomical Journal* 68 (September 1963): 515–521.

2. Peter van de Kamp to Gerard Kuiper, January 18, 1940, Astronomy Department Records, SCA.

3. Peter van de Kamp to Kaj Strand, June 11, 1953, Astronomy Department Records, SCA.

4. Edwin Dennison to Peter van de Kamp, February 22, 1955, Astronomy Department Records, SCA.
5. Peter van de Kamp to John Lear, science editor, *Saturday Review*, May 22, 1962, Astronomy Department Records, SCA.
6. Carle Hodge, "Scientist Reports Planet Without Seeing It," *Arizona Daily Sun*, April 19, 1963.
7. "New Neighbor," *Newsweek*, April 29, 1963.
8. Peter van de Kamp to Lawrence J. Tacker, Major, USAF, August 25, 1959, Astronomy Department Records, SCA.
9. Brown, "Scientists Search for Signs of Life."
10. Su-Shu Huang, "The Problem of Life in the Universe and the Mode of Star Formation," *Publications of the Astronomical Society of the Pacific* 71, no. 422 (1959).
11. Science Service, "Planet Discovered Outside of the Sun's Solar System," *Harrisonburg (VA) News Record*, April 25, 1963.
12. "New Neighbor."
13. "Another Solar System Is Found 36 Trillion Miles from the Sun," *New York Times*, April 19, 1963, 4.
14. "Invisible Planet Is Discovered by Sproul Director," *Philadelphia Inquirer*, April 19, 1963.
15. Edwin Hubble, "First Photographs with the 200-inch Hale Telescope," *Publications of the Astronomical Society of the Pacific* 61, no. 360 (June 1949): 121.
16. "Super-Prof Wins National Fame," *Swarthmore College Phoenix*, September 20, 1966.
17. Walter Sullivan, *We Are Not Alone* (New York: McGraw-Hill, 1966), 49.
18. Sullivan, *We Are Not Alone*, 52.
19. Sullivan, *We Are Not Alone*, 53.
20. "The Invisible Planet," *Experiment: The Story of a Scientific Search*, produced by Prism Productions, aired July 1966 on National Educational Television.
21. Peter van de Kamp, "The Philosophical Man" (speech at the *Man and the Twenty-First Century* conference, Stevens Institute of Technology, Hoboken, New Jersey, November 14, 1967).
22. Claire Huff, "Scientist Skeptical on Space Flights," *Philadelphia Inquirer*, January 14, 1968.

23. "Planets in His Pockets, He Laughs about UFOs," *Montreal Star*, November 15, 1968.

24. Jeff Yohay, "Student Interviews Dr. van de Kamp; Derides Vietnam Effort, Space Race" (Stevens Institute Centennial Convocation, November 17, 1967).

25. Peter van de Kamp, "Tribute to Laika," collection speech, Clothier Memorial, December 5, 1957.

26. Van de Kamp, "Tribute to Laika."

27. Rosalie Piersol, "New Wing 'Shot in Arm,'" Philadelphia *Sunday Bulletin*, October 27, 1963, 5.

28. Nancy G. Roman interview by David DeVorkin, August 19, 1980, Oral Histories, Niels Bohr Library and Archives, American Institute of Physics, http://www.aip.org/history-programs/niels-bohr-library/oral-histories/4846.

29. Daniel Hoffman, "The Peaceable Kingdom" (poem written for Phi Beta Kappa Honor Society, Swarthmore, Pennsylvania, June 7, 1964).

30. "NSF Gives $49,900 for Star Studies," *Swarthmore College Phoenix*, January 8, 1965.

31. "Faculty Members Gain Promotions for Upcoming Year," *Swarthmore College Phoenix*, April 16, 1965.

32. Peter van de Kamp, "Parallax and Proper Motion of BD +5 1668," *Astronomical Journal* 53 (August 1948): 229.

33. Ann Ewing, "Unseen Companion," *Science News* 89, no. 296 (April 23, 1966): 296.

34. NASA, "How Many Exoplanets Has Kepler Discovered?" NASA.gov, accessed January 5, 2018, https://www.nasa.gov/kepler/discoveries.

35. James F. Wanner, *The 24-inch Refractor of the Sproul Observatory* (Swarthmore, PA: Swarthmore College, 1972), 3; originally published as "New Life for Old Sproul Refractor," in *Popular Astronomy* 62, no. 550 (1968): 11–15.

36. Peter van de Kamp to Kaj Aa. Strand, October 24, 1969, Astronomy Department Records, SCA.

37. Elizabeth Weber, "The Crisis of 1969," Swarthmore *Phoenix*, March 7, 1996.

38. Courtney Smith, "The Academic Community and Social Concerns," *Swarthmore College Bulletin*, December 1965.

39. James Wanner to Peter van de Kamp, February 18, 1964, Astronomy Department Records, SCA.

40. Paul Good, "Requiem for Courtney Smith," *Life*, May 9, 1969, 76.

41. Stuart Madden, "Swarthmore President's Death Ends Black Sit-In," *Daily Pennsylvanian*, January 17, 1969.

42. "Dr. Courtney C. Smith, Swarthmore's President," *New York Times*, January 17, 1969, 44.

43. Good, "Requiem for Courtney Smith," 76D.

44. Etheridge quoted in Good, "Requiem for Courtney Smith," 83.

45. Darwin H. Stapleton and Donna Heckman Stapleton, *Dignity, Discourse, and Destiny: The Life of Courtney C. Smith* (Newark: University of Delaware Press, 2004), 186.

46. Peter van de Kamp, "Based on what I said at the Chaplin Seminar February 1," January 1969, Astronomy Department Records, SCA.

47. "Subversion of Reason," *New York Times*, January 18, 1969, 30.

Chapter Four

1. Alisa Giardinelli, "Longtime Astronomy Professor Wulff Heintz Dies," *Swarthmore News Archive 2006–2007*, June 16, 2006, https://www.swarthmore.edu/news-archive-2006-2007/longtime-astronomy-professor-wulff-heintz-dies.

2. Harry J. Augensen and Edward H. Geyer, "Wulff-Dieter Heintz (1930–2006)," American Astronomical Society Obituaries, June 2006, https://aas.org/obituaries/wulff-dieter-heintz-1930-2006.

3. Peter van de Kamp, "Alternate Dynamical Analysis of Barnard's Star," *Astronomical Journal* 74 (April 18, 1969): 757–759.

4. Shiv S. Kumar to Peter van de Kamp, March 21, 1969, Astronomy Department Records, SCA.

5. Sandra Faber to Peter van de Kamp, May 12, 1969, Astronomy Department Records, SCA.

6. Martin Putnam, "Van de Kamp Discovers New Planetary System," *Swarthmore Phoenix*, April 22, 1969.

7. "The Crab Pulsar: Many New Observations," IAU Symposium No. 46, *Assembly Times: The Daily Bulletin of the 14th General Assembly of the International Astronomical Union*, August 21, 1970.

8. Peter van de Kamp to Kaj Aa. Strand, September 21, 1971, Astronomy Department Records, SCA.

9. "The Star-Planet," *Time*, January 29, 1973.

10. Peter van de Kamp, "Parallax and Orbital Motion of Epsilon Eridani," *Astronomical Journal* 79, no. 4 (1974): 491–492.

11. Jill Tarter, "Brown Dwarfs and Black Holes," *Astronomy*, April 1978, 24.

12. Viki Joergens, ed., *50 Years of Brown Dwarfs* (Cham: Springer, 2013), 20.

13. Walter Sullivan, "Is Jupiter a Star?" *New York Times*, December 19, 1965.

14. Sarah Lee Lippincott, "Reflections on 24" Refractor," *Swarthmorean*, March 8, 1974.

15. Peter van de Kamp and Michael D. Worth, "Parallax and Orbital Motion of the Unresolved Astrometric Binary BD+43 4305," *Astronomical Journal* 77, no. 9 (November 1972): 762–763.

16. Peter van de Kamp and Sarah Lee Lippincott, "Continued Astronometric Study of BD +43 4305" (paper presented at the 156th American Astronomical Society Meeting, June 15–18, 1980, College Park, Maryland).

17. Sarah Lee Lippincott to Edwin Dennison, November 25, 1974, Astronomy Department Records, SCA.

18. Teresa Nicholas, "Van de Kamp Offers Fourth Chaplin Seminar, Expresses Advantages of Old Motion Pictures," *Swarthmore Phoenix*, April 29, 1975.

19. A. R. Martin, *Project Daedalus: The Final Report on the BIS Starship Study* (London: British Interplanetary Society, 1978).

Chapter Five

1. Peter van de Kamp to Edward Litchfield, February 1, 1965, Astronomy Department Records, SCA.

2. "Crisis at Allegheny," *Sky and Telescope* 29, no. 3 (March 1965): 135.

3. "Pitt Phases Out Observatory Program," press release, January 21, 1964, Pitt Administration Collection, University Library System Digital Collections, University of Pittsburgh, http://digital.library.pitt.edu/islandora/objectpitt:pittpressreleases19640018/from_search/5bc49877965cc08ddcbe58010e00cb42-3#page/1/mode/2up.

4. "Academic Review: Division of the Natural Sciences," *Pitt* 23, no. 4 (Fall 1967): 13, Pitt Alumni Collection, University Library System Digital Collections, University of Pittsburgh, http://digital.library.pitt.edu/islandora/object/pitt:31735062134451/from_search/bbb2b24b00b4629e567fab0b09a11f22-6#page/18/mode/2up.

5. Olivier Jensen and Tadeusz Ulrych, "An Analysis of the Perturbations on Barnard's Star," *Astronomical Journal* 78 (December 1973): 1104.

6. Olivier Jensen and Tadeusz Ulrych, "Erratum: An Analysis of the Perturbations on Barnard's Star [Astron. J. 78, 1104 (1973)]," *Astronomical Journal* 79 (November 1974): 1328.

7. George Gatewood and Heinrich Eichhorn, "An Unsuccessful Search for a Planetary Companion of Barnard's star BD +4 3561," *Astronomical Journal* 78 (October 1973): 769–776.

8. John L. Hershey, "Astrometric Analysis of the Field of AC +65 6955 from Plates Taken with the Sproul 24-inch Refractor," *Astronomical Journal* 78 (June 1973): 421–425.

9. H. K. Eichhorn von Wurmb to Swarthmore, May 16, 1975, Astronomy Department Records, SCA.

10. "Report of the Visiting Committee."

11. Sarah Lee Lippincott, "Technical Report to the National Science Foundation," February 25, 1979, Astronomy Department Records, SCA.

12. Strand interview, December 8, 1983.

13. George Gatewood, "An Astrometric Study of Lalande 21185," *Astronomical Journal* 79, no. 1 (January 1974): 52–53.

14. George Gatewood, "On the Astrometric Detection of Neighboring Planetary Systems," *Icarus* 27, no. 1 (January 1976): 1–12.

15. Jane Russell, George Gatewood, and N. E. Wagman, "Altair," *Astronomical Journal* 83, no. 11 (November 1978): 1458.

16. George Gatewood, Lee Breakiron, Ronald Goebel, Steven Kipp, Jane Russell, and John Stein, "On the Astrometric Detection of Neighboring Planetary Systems, II," *Icarus* 41, no. 2 (February 1980): 205–231.

17. Wulff-Dieter Heintz, "Reexamination of Suspected Unresolved Binaries," *Astrophysical Journal* 220 (March 15, 1978): 931–934.

18. Sarah Lee Lippincott and Wulff-Dieter Heintz, "Swarthmore College: Sproul Observatory and Department of Astronomy, Swarthmore, Pennsylvania 19081, Report for the Period 1 July 1977–30 June 1978," *Bulletin of the Astronomical Society* 11 (January 1979): 315–316.

19. John L. Hershey and Elliot Borgman, "Upper Limits on the Mass of a Dark Companion of Groombridge 1618 from the 40-Year Sproul Plate Series," *Bulletin of the American Astronomical Society* 10 (September 1978): 630.

20. Sarah Lee Lippincott and Wulff-Dieter Heintz, "Sproul Observatory," *Bulletin of the American Astronomical Society* 12 (January 1980): 367.

21. Bill Kent, "Barnard's Wobble," *Swarthmore Bulletin*, March 2001, 31.

22. Van de Kamp interview, March 18, 1979.

23. Van de Kamp interview, March 18, 1979; Strand interview, December 8, 1983.

24. Phillip J. Ianna, "On Aberrations and Field Errors," *Vistas in Astronomy* 6, no. 1 (1965): 93–123; and Phillip J. Ianna, "Aberrations and Field Errors of the Sproul 24-inch Objective," *Astronomical Journal* 67 (June 1962): 273.

25. Minutes, NASA Ames Astrometric Conference, US Naval Observatory, Washington, DC, May 10–11, 1976, Astronomy Department Records, SCA.

26. L. W. Fredrick and P. A. Ianna, "A Study of Barnard's Star," *Bulletin of the American Astronomical Society* 12 (March 1980): 455.

27. L. W. Fredrick and P. A. Ianna, "The Barnard's Star Perturbation," *Bulletin of the American Astronomical Society* 17 (March 1985): 551.

28. Peter van de Kamp, "Barnard's Star 1916–1976: A Sexagintennial Report," *Vistas in Astronomy* 20, no. 2 (1977): 501–521.

29. Peter van de Kamp, "Sproul Astrometric Study of Barnard's Star, A Progress Report," *Bulletin of the American Astronomical Society* 14 (March 1982): 627.

30. Peter van de Kamp, "Friedrich Wilhelm Bessel 1784, July 22–1846, March 17," *Astrophysics and Space Science* 110, no. 1 (March 1985): 103–104.

31. A. H. Batten, "Book Review: Stellar Paths," *Journal of the Royal Astronomical Society of Canada* 77 (February 1983): 59–60.

32. Peter van de Kamp, "Dark Companions of Stars: Astrometric Commentary on the Lower End of the Main Sequence," *Space Science Reviews* 43 (April 1986): 211–327. NSF-supported research.

Chapter Six

1. Sarah Lee Lippincott and Michael D. Worth, "Chi-1 Orionis, a New Solar-Type Astrometric Binary," *Publications of the Astronomical Society of the Pacific* 90, no. 535 (June 1978): 330–332.

2. G. F. Benedict, B. E. McArthur, O. G. Franz, L. H. Wasserman, T. J. Henry, T. Takato, I. V. Strateva, et al., "Precise Masses for Wolf 1062 AB from Hubble Space Telescope Interferometric Astrometry and McDonald Observatory Radial Velocities," *Astronomical Journal* 121, no. 3 (March 2001): 1607–1613.

3. Jennifer Burt, Steven S. Vogt, R. Paul Butler, Russell Hanson, Stefano Meschiari, Eugenio J. Rivera, Gregory W. Henry, and Gregory Laughlin, "The Lick-Carnegie

Exoplanet Survey: Gliese 687b—A Neptune-Mass Planet Orbiting a Nearby Red Dwarf," *Astrophysical Journal* 789, no. 2 (2014), https://doi.org/10.1088/0004-637X/789/2/114.

4. Sarah Lee Lippincott, "Astrometric Analyses of the Unseen Companions to Ci 18, 2354 and Wolf 1062 from Plates Taken with the 61-cm Sproul Refractor," *Astronomical Journal* 82 (November 1977): 925–928.
5. Sarah Lee Lippincott to Laurence W. Fredrick, November 28, 1977, Astronomy Department Records, SCA.
6. Sarah Lee Lippincott, "Two New Astrometric Binaries of Small Amplitude: One with an Unseen Member of ~20 Jupiter Masses," *Bulletin of the American Astronomical Society* 11 (1979): 405.
7. Hershey and Borgman, "Upper Limits on the Mass of a Dark Companion," 630.
8. Geoff Marcy and David Moore, "The Extremely Low Mass Companion to Gliese 623," *Astrophysical Journal* 341, part 1 (June 15, 1989): 961–967.
9. N. A. Shakht, "The Observations of Gliese 623 and Some other Objects with Suspected Unseen Components," in *Astronomical and Astrophysical Objectives of Sub-Milliarcsecond Optical Astrometry*, ed. Erick Høg, 359 (Dordrecht: Springer Netherlands, 1995).
10. Frantz Martinache, James P. Lloyd, Michael J. Ireland, Ryan S. Yamada, and Peter G. Tuthill, "Precision Masses of the Low-Mass Binary System GJ 623," *Astrophysical Journal* 661, no. 1 (2007): 496–501.
11. Sarah Lee Lippincott to Laurence W. Fredrick, June 5, 1980, Astronomy Department Records, SCA.
12. J.-L. Beuzit, D. Ségransan, T. Forveille, S. Udry, X. Delfosse, M. Mayor, C. Perrier, et al., "New Neighbours, III: 21 New Companions to Nearby Dwarfs, Discovered with Adaptive Optics," *Astronomy and Astrophysics* 425, no. 3 (October 3, 2004): 997–1008.
13. Sarah Lee Lippincott and John L. Hershey, "Letter of Intent in Answer to the Announcement of Opportunity for Space Telescope: Search for Faint Companions to Nearby Stars—an Extension of the Sproul Astrometric Program," April 22, 1977, Astronomy Department Records, SCA.
14. G. Westerhout, H. Eichhorn, G. Gatewood, J. Hughes, W. Jefferys, I. King, and W. van Altena, "Astronomy Survey Committee," Astrometry Working Group, third draft, February 1980, SCA.
15. A. B. Schultz, H. M. Hart, J. L. Hershey, F. C. Hamilton, M. Kochte, F. C. Bruhweiler, G. F. Benedict, et al., "A Possible Companion to Proxima Centauri," *Astronomical Journal* 115 (January 1998): 345.

16. Gabrielle Walker, "Did Hubble Catch a Glimpse of a Nearby Giant?" *New Scientist*, no. 2119 (January 31, 1998), https://www.newscientist.com/article/mg15721190-700-did-hubble-catch-a-glimpse-of-a-nearby-giant/.

17. Guillem Anglada-Escudé, Pedro J. Amado, John Barnes, Zaira M. Berdiñas, R. Paul Butler, Gavin A. L. Coleman, Ignacio de la Cueva, et al., "A Terrestrial Planet Candidate in a Temperate Orbit around Proxima Centauri," *Nature* 536 (August 25, 2016): 437–440

18. Guillem Anglada, Pedro J. Amado, José L. Ortiz, José F. Gómez, Enrique Macías, Antxon Alberdi, Mayra Osorio, et al.. "ALMA Discovery of Dust Belts around Proxima Centauri," *Astrophysical Journal Letters* 850, no. 1 (November 15, 2017): 1–5.

19. John Wenz, "Proxima Centauri's Dust Belt Hints at More Planets," *Discover*, November 3, 2017, http://blogs.discovermagazine.com/d-brief/2017/11/03/alpha-centauri-system.

20. Sarah Lee Lippincott, "Continued Astrometric Study of BD+43^{0}4305," *Bulletin of the American Astronomical Society* 12 (March 1980): 455.

21. Sarah Lee Lippincott, "EV Lacertae—Is Flare Activity Related to an Unseen Planet-like Companion?" *Activity in Red-Dwarf Stars: Proceedings of the 71st Colloquium of the International Astronomical Union, Held in Catania, Italy, August 10–13, 1982*, ed. Patrick B. Byrne and Marcello Rodonò (Dordrecht: Springer Netherlands, 1983), 201–202.

22. Kaj Strand, "Triple System Stein 2051 (G 175–34)," *Astronomical Journal* 82 (September 1977): 745–749.

23. Peter van de Kamp to Kaj Strand, September 18, 1974, Astronomy Department Records, SCA.

24. Kaj Strand to Peter van de Kamp, October 8, 1974, Astronomy Department Records, SCA.

25. Mark Bowden, "Stargazers, Dancers and Lifelong Searches," *Philadelphia Inquirer*, September 26, 1980.

26. John Hershey to David C. Black, May 8, 1980, Astronomy Department Records, SCA.

27. K. A. Strand and V. V. Kallarkal, "The Nearby Binary, Stein 2051 (G175–34AB)," in *White Dwarfs*, ed. G. Wegner, Lecture Notes in Physics, vol. 328 (Berlin: Springer, 1989).

28. Kailash Sahu, C. Anderson, Jay Casertano, Stefano Bond, Howard E. Bergeron, Pierre Nelan, Edmund P. Pueyo, et al., "Relativistic Deflection of Background Starlight Measures the Mass of a Nearby White Dwarf Star," *Science* 356, no. 6342 (June 2017): 1046–1050.

29. Burt et al., "The Lick-Carnegie Exoplanet Survey," 114.

30. George Gatewood, "MAP Determinations of the Parallaxes of Stars in the Regions of HD 2665, BD +68.946 deg, and Lambda Ophiuchi," *Astronomical Journal* 97 (April 1989): 1189–1196.

31. Fred Burwell, "Fridays with Fred: The Science behind Smith Observatory," *Beloit News*, March 24, 2011, https://www.beloit.edu/campus/news/fwf/?story_id=313226.

32. Linda Scott, "Eye on the Sky," *Abington Suburban*, July 7, 2016.

33. Faber interview, October 15, 1988.

34. The *National Enquirer* article is incomplete in the Swarthmore College Archives, Astronomy Department Records.

35. Sally Bedell, "Dave Garroway, 69, Found Dead; First Host of 'Today' on NBC-TV," *New York Times*, July 22, 1982.

36. Associated Press, "Dave Garroway Family Plans a Jazz Concert as Memorial," *New York Times*, July 25, 1982.

37. Kenneth Clark, "Friends, Colleagues Mourn Dave Garroway," United Press International, July 21, 1982.

38. Sarah Lee Lippincott, "An Unseen Companion to 36 Ursae Majoris a From Analysis of Plates taken with the Sproul 61-cm Refractor," *Astronomical Society of the Pacific* 95 (October 1983): 775–777.

Chapter Seven

1. David Black, ed., *Project Orion: A Design Study of a System for Detecting Extrasolar Planets*, NASA SP 436 (Washington, DC: Government Printing Office, 1981).

2. George D. Gatewood, *The Allegheny Observatory Search for Planetary Systems: Final Technical Report, 1980–1988* (Department of Physics and Astronomy, University of Pittsburgh, Pennsylvania, March 1989).

3. G. Gatewood, M. W. Castelaz, J. W. Stein, E. H. Levy, R. S. McMillan, K. Nishioka, and J. D. Scargle, "A Prototype Detector for the Astrometric Telescope Facility," in *Mapping the Sky: Past Heritage and Future Directions, Proceedings of the 133rd Symposium of the International Astronomical Union, Held in Paris, France, June 1–5, 1987*, ed. S. Debarbat, J. A. Eddy, H. K. Eichhorn, and A. R. Upgren (Dordrecht: Kluwer, 1988), 421–424.

4. Jean Kovalevsky, *Modern Astrometry* (Berlin: Springer, 2013), 120.

5. George D. Gatewood, *Allegheny Observatory Search for Planetary Systems*, Reports of Planetary Astronomy, (Washington, DC: NASA, 1991), 45–46.

6. George D. Gatewood, "A Study of the Astrometric Motion of Barnard's Star," *Astrophysics and Space Science* 223, no. 1–2 (1995): 91–98.

7. G. Fritz Benedict, Barbara McArthur, D. W. Chappell, E. Nelan, W. H. Jefferys, W. van Altena, J. Lee, et al., "Interferometric Astrometry of Proxima Centauri and Barnard's Star Using Hubble Space Telescope Fine Guidance Sensor 3: Detection Limits for Substellar Companions," *Astronomical Journal* 118, no. 2 (1999): 1086–1100.

8. M. Kürster, M. Endl, F. Rouesnel, S. Els, A. Kaufer, S. Brillant, A. P. Hatzes, S. H. Saar, and W. D. Cochran, "The Low-Level Radial Velocity Variability in Barnard's Star (= GJ 699)," *Astronomy and Astrophysics* 403, no. 3 (June 1, 2003): 1077–1087.

9. Wulff-Dieter Heintz, "The Substellar Masses of Wolf 424," *Astronomy and Astrophysics* 217, no. 1–2 (1989): 145.

10. Michael Decourcy Hinds, "Swarthmore Journal: Life's Quest Rewarded, 84 Trillion Miles Away," *New York Times*, September 28, 1989.

11. Wulff-Dieter Heintz, "The Low-Mass Binary Hei 299," *Observatory* 110 (August 1990): 131–132.

12. W. L. Combrinck, "Astrometry—Extrasolar Planet Detection Method," *Monthly Notes of the Astronomical Society of Southern Africa* 42 (1983): 89.

13. Thomas O'Toole, "Possibly as Large as Jupiter," *Washington Post*, December 31, 1983.

14. Carl Sagan, Frank Drake, Ann Druyan, Timothy Ferris, Jon Lomberg, and Linda Salzman Sagan, *Murmurs of Earth* (New York: Random House, 1978), 235.

15. Nathan Kaib, Ethan White, and Andre Izidoro,"Fomalhaut's Stellar Companions as the Driver of Its Morphology" (talk at the American Astronomical Society Meeting 231, January 2018).

16. Alex Teachey, David M. Kipping, and Allan R. Schmitt, "HEK VI: On the Dearth of Galilean Analogs in Kepler and the Exomoon Candidate Kepler-1625b I," *Astronomical Journal* 155, no. 1 (December 22, 2017): 1–20.

17. A. Lecavelier des Etangs and A. Vidal-Madjar, "Is Beta Pic b the Transiting Planet of November 1981?" *Astronomy and Astrophysics* 497, no. 2 (April 2009): 557–562.

18. A.-M. Lagrange, A. Boccaletti, M. Langlois, G. Chauvin, R. Gratton, H. Beust, S. Desidera, et al., "Post Conjunction Detection of Beta Pictoris b with VLT/SPHERE," preprint, submitted September 22, 2018, https://arxiv.org/abs/1809.08354.

19. Rafael Rebolo, "Teide 1 and the Discovery of the Brown Dwarfs," in *50 Years of Brown Dwarfs*, ed. Viki Joergens (Cham: Springer, 2013), 27.

20. R. S. Harrington, V. V. Kallarakal, and C. C. Dahn, "Astrometry of the Low-Luminosity Stars VB8 and VB10," *Astronomical Journal* 88 (July 1983): 1038–1039.

21. John Noble Wilford, "Possible Planet Found Beyond the Solar System," *New York Times*, December 11, 1984.

22. John Noble Wilford, "Scientists Dispute Planet Discovery," *New York Times*, October 5, 1986.

23. John Noble Wilford, "Orbiting Object Hints at Other Planet Systems," *New York Times*, November 11, 1987.

24. Jacob L. Bean, Andreas Seifahrt, Henrik Hartman, Hampus Nilsson, Ansgar Reiners, Stefan Dreizler, Todd J. Henry, and Günter Wiedemann, "The Proposed Giant Planet Orbiting VB 10 Does Not Exist," *Astrophysical Journal Letters* 711, no 1 (February 11, 2010): L19–L23.

25. Michael Perryman, Joel Hartman, Gaspar A Bakos, and Lennard Lindegren, "Astrometric Exoplanet Detection with Gaia," *Astrophysical Journal* 797, no. 1 (November 19, 2014): 1–22.

26. Solar System Exploration Committee, NASA Advisory Council, *Planetary Exploration Through Year 2000: An Augmented Program* (Washington, DC: Government Printing Office, 1986).

27. "1917 Astronomical Plate Has First-Ever Evidence of Exoplanetary System," Carnegie Science, April 12, 2016, https://carnegiescience.edu/news/1917-astronomical-plate-has-first-ever-evidence-exoplanetary-system.

28. Solar System Exploration Committee, *Planetary Exploration Through Year 2000*, 191.

29. W. J., Borucki, J. D. Scargle, and H. S. Hudson, "Detectability of Extrasolar Planetary Transits," *Astrophysical Journal* 291 (April 15, 1985): 852–854.

30. Bernard Burke, "Study of Extrasolar Planetary Systems from the Moon: Summary of the Panel Discussion," *AIP Conference Proceedings* 207, no. 1 (1990): 67, https://doi.org/10.1063/1.39358.

31. Bruce Campbell, G. A. H. Walker, and S. Yang, "A Search for Substellar Companions to Solar-Type Stars," *Astrophysical Journal* 331, no. 1 (Aug. 15, 1988): 902–921.

32. Paul Butler, "A Brief Personal History of Exoplanets," Carnegie Science, March 29, 2016, https://dtm.carnegiescience.edu/news/brief-personal-history-exoplanets.

33. David W. Latham, Tsevi Mazeh, Robert P. Stefanik, Michel Mayor, and Gilbert Burki, "The Unseen Companion of HD114762: A Probable Brown Dwarf," *Nature* 339 (May 4, 1989): 38–40.

34. W. J. Forrest, J. D. Garnett, Z. Ninkov, M. Skrutskie, and M. Shure, "Discovery of Low Mass Brown Dwarfs in Taurus," *Bulletin of the American Astronomical Society* 21 (March 1989): 795.

35. Ben R. Oppenheimer, "Companions of Stars," in *50 Years of Brown Dwarfs*, ed. Viki Joergens (Cham: Springer, 2013), 96.

Chapter Eight

1. M. Demiański and M. Prószyński, "Does PSR0329 + 54 Have Companions?" *Nature* 282 (November 22, 1979): 383–385.

2. T. V. Shabanova, "Evidence for a Planet around the Pulsar PSR B0329+54," *Astrophysical Journal* 453 (November 1995): 779.

3. B. Cadwell, "Pulsar Survey and Timing with the Penn State Pulsar Machines," *Bulletin of the American Astronomical Society* 29, no. 5 (December 1997): 1396.

4. E. D. Starovoit and A. E. Rodin, "On the Existence of Planets around the Pulsar PSR B0329+54," *Astronomy Reports* 61, no. 11 (November 2017): 948–953.

5. A. Wolszczan and D. A. Frail, "A Planetary System around the Millisecond Pulsar PSR1257 + 12," *Nature* 355 (Jan. 1992): 145–147.

6. A. Patruno and M. Kama, "Neutron Star Planets: Atmospheric Processes and Irradiation," *Astronomy and Astrophysics* 608 (2017): A147.

7. Michael D. Lemonick, *Other Worlds: The Search for Life in the Universe* (New York: Simon and Schuster, 1998), 87.

8. Doug Braun, "Barnard's Star and Planets?" Astro.sci Usenet group, March 23, 1995.

9. Sarah Lee Lippincott, "Peter van de Kamp, 1901–1995," *Bulletin of the American Astronomical Society* 27 (December 1995): 1483–1484.

10. David Stout, "Peter van de Kamp, Astronomer and Musician at Swarthmore, 93," *New York Times*, May 23, 1995.

11. Laurence Fredrick, "Peter van de Kamp (1901–1995)," *Publications of the Astronomical Society of the Pacific* 108 (August 1996): 556.

12. Marcus Woo, "The First Planet Ever Discovered around Another Star," *BBC Earth*, October 20, 2016, http://www.bbc.com/earth/story/20161019-the-first-planet-around-another-star.

13. Michel Mayor and Didier Queloz, "A Jupiter-Mass Companion to a Solar-Type Star," *Nature* 378, no. 6555 (1995): 355–359.

14. Rebolo, "Teide 1 and the Discovery of Brown Dwarfs," 25–47.

15. Maggie A. Thompson, J. Davy Kirkpatrick, Gregory N. Mace, Michael C. Cushing, Christopher R. Gelino, Roger L. Griffith, Michael F. Skrutskie, Peter R. M. Eisenhardt, Edward L. Wright, and Kenneth A. Marsh, "Nearby M, L, and T Dwarfs Discovered by the Wide-field Infrared Survey Explorer (WISE)," *Publications of the Astronomical Society of the Pacific* 125, no. 929 (July 1 2013): 809–837.

16. George Gatewood, "Lalande 21185," *Bulletin of the American Astronomical Society* 28 (May 1996): 885.

17. Natalia Shakht, "A Study of the Motion of the Nearby Red Dwarf Lalande 21185," in "Complementary Approaches to Double and Multiple Star Research," ed. H. A. McAlister and W. I. Hartkopf, *International Astronomical Union Colloquium* 135 (1992): 346.

18. George Gatewood, John Stein, Joost K. de Jonge, Timothy Persinger, Thomas Reiland, and Bruce Stephenson, "Multichannel Astrometric Photometer and Photographic Astrometric Studies in the Regions of Lalande 21185, BD 56 deg 2966, and HR 4784," *Astronomical Journal* 104, no. 3 (September 1992): 1237–1247. Research supported by University of Pittsburgh and Allegheny Observatory.

19. D. W. McCarthy Jr., "Near Infrared Imaging of Unseen Companions to Nearby Stars," in "The Nearby Stars and the Stellar Luminosity Function," ed. A.G.D. Philip and A. R. Upgren, *International Astronomical Union* Colloquium 76 (1983): 107.

20. Tommy Ehrbar, "Invisible Dancer," *Pitt Magazine*, January 1997, https://web.archive.org/web/20160303174636/https://www.pittmag.pitt.edu/jan97/gatewoodh.html.

21. Virginia Trimble and Lucy-Ann McFadden, "Astrophysics in 1997," *Publications of the Astronomical Society of the Pacific* 110 no. 745 (March 3, 1998): 224.

22. Bruce Dorminey, *Distant Wanderers: The Search for Planets Beyond the Solar System* (New York: Copernicus, 2002), 116.

23. Natalia Shakht, "The History of the Search for Extrasolar Planetary Systems at Pulkovo Observatory," in *Planetary Systems: The Long View*, ed. L. M. Celnikier and J. Trân Thanh Van (Gif-sur-Yvette, France: Editions Frontières, 1998), 395.

24. Alycia J. Weinberger, Geoff Bryden, Grant M. Kennedy, Aki Roberge, Denis Defrère, Philip M. Hinz, Rafael Millan-Gabet, et al., "Target Selection for the LBTI Exozodi Key Science Program," *Astrophysical Journal Supplement* 216, no. 2 (January 27, 2015): 24.

25. R. Paul Butler, Steven S. Vogt, Gregory Laughlin, Jennifer A. Burt, Eugenio J. Rivera, Mikko Tuomi, Johanna Teske, et al., "The LCES HIRES/Keck Precision Radial Velocity Exoplanet Survey," *Astronomical Journal* 153, no. 5 (April 13, 2017): 1–19.

26. Abel Mendez, "A New Search for Extrasolar Planets from the Arecibo Observatory," press release, Planetary Habitability Laboratory, July 12, 2017, http://phl.upr.edu/press-releases/barnard.

27. G. Fritz Benedict, Barbara E. McArthur, George Gatewood, Edmund Nelan, William D. Cochran, Artie Hatzes, Michael Endl, et al., "The Extrasolar Planet ε Eridani b: Orbit and Mass," *Astronomical Journal* 132, no. 5 (November 2006): 2206–2218.

28. Joshua Roth, "Does 51 Pegasi's Planet Really Exist?" *Sky and Telescope* 93, no. 5 (May 1997): 24–27.

29. G. A. H.Walker, D. A. Bohlender, A. R.Walker, A. W. Irwin, S. L. S. Yang, and A. Larson, "Gamma Cephei—Rotation or Planetary Companion?," *Astrophysical Journal* 396, no. 2, part 2—letters (September 10, 1992): L91–L94.; Jacob Berkowitz, "Lost World: How Canada Missed Its Moment of Glory," *Globe and Mail*, May 1, 2018.

30. Artie P. Hatzes, William D. Cochran, Michael Endl, Barbara McArthur, Diane B. Paulson, Gordon A. H. Walker, Bruce Campbell, and Stephenson Yang, "A Planetary Companion to γ Cephei A," *Astrophysical Journal* 599, no. 2 (December 2003): 1383–1394.

31. Stephen R. Kane and Dawn M. Gelino, "Distinguishing between Stellar and Planetary Companions with Phase Monitoring," *Monthly Notices of the Royal Astronomical Society* 424, no. 1 (July 21, 2012): 779–788.

Chapter Nine

1. Ben Guarino, "The Hottest Planet Ever Discovered Has an Atmosphere as Warm as a Star's," *Washington Post*, June 5, 2017, https://www.washingtonpost.com/news/speaking-of-science/wp/2017/06/05/the-hottest-planet-ever-discovered-has-an-atmosphere-as-warm-as-a-stars.

2. I. Ribas, M. Tuomi, A. Reiners, R. P. Butler, J. C. Morales, M. Perger, S. Dreizler, et al., "A Super-Earth Planet Candidate Orbiting at the Snow-Line of Barnard's Star," *Nature* 563 (November 15, 2018): 365–368. I reviewed a preprint made available by the authors.

3. Daniel Bayliss, Edward Gillen, Philipp Eigmuller, James McCormac, Richard D. Alexander, David J. Armstrong, Rachel S. Booth, et al., "NGTS-1b: A Hot Jupiter Transiting an M-Dwarf," *Monthly Notices of the Royal Astronomical Society* 475, no. 4 (April 2018): 4467–4475.

4. Mikko Tuomi, Hugh R. A. Jones, John R. Barnes, Guillem Anglada-Escudé, and James S. Jenkins, "Bayesian Search for Low-Mass Planets around Nearby M Dwarfs: Estimates for Occurrence Rate Based on Global Detectability Statistics," *Monthly Notices of the Royal Astronomical Society* 441, no. 2 (June 21, 2014): 1545–1569.

Index

Page numbers in italic refer to figures.